SEEING BEYOND
THE VISIBLE

Contributors

A. HEWISH   M.A., Ph.D., F.R.S.
Mullard Radio Astronomy Observatory
University of Cambridge

M. W. OVENDEN   M.A., Ph.D.
Professor of Astronomy
University of British Columbia

DAME KATHLEEN LONSDALE   D.B.E., D.Sc., LL.D., F.R.S.
Department of Chemistry
University College London

JUDITH MILLEDGE   M.Sc., D.Sc.
Department of Chemistry
University College London

A. L. CULLEN   O.B.E., D.Sc.
Head of Department of Electronic
and Electrical Engineering
University College London

D. J. E. INGRAM   M.A., D.Phil., D.Sc.
Professor of Physics
University of Keele

SEEING BEYOND
THE VISIBLE

Edited by A. Hewish

THE ENGLISH UNIVERSITIES PRESS LTD
ST PAUL'S HOUSE WARWICK LANE LONDON EC4

ISBN 0 340 09894 5

First printed 1970

Printed in Great Britain for The English Universities Press Ltd by
Richard Clay (The Chaucer Press), Ltd, Bungay, Suffolk

EDITOR'S FOREWORD

The essays contained in this book originated as a series of talks broadcast by the B.B.C. under the title 'At the Speed of Light'. The choice of topics and the mode of treatment owe much to the producer, Rosemary Jellis, and to Prof. F. Graham Smith, who collaborated with her and provided a commentary throughout the series. In their essays, however, the authors have gone considerably further than was possible within the compass of the original talks.

The electromagnetic spectrum embraces a wide range of phenomena between radio wavelengths at one end and gamma radiation at the other. It is one of the major achievements of physics that all of these may be understood in terms of a single theory. Modern scientific instruments now enable us to extend the limited capabilities of the human eye so that we can 'see' over far broader horizons of size and distance, both outwards to the Universe and inwards towards the atom. The exciting new knowledge gained from this exploitation of the electromagnetic spectrum provides the theme of this book.

Chapters One and Two are introductory and provide a background which is necessary in order to understand the wide variety of topics discussed. These show how the theory developed from the early notions of the Greeks to the sophisticated wave-particle concepts of the present day. The approach is entirely non-mathematical, the aim being to present the important physical ideas and to show how they came into being as a result of careful observations of the behaviour of radiation and its interaction with matter.

The atmosphere surrounding the earth screens off most of the radiation arriving from outer space, but there are two 'windows', one of which lets through visible light and the other radio waves. Chapters Three and Four describe the use that is made of these windows by optical and radio telescopes. It is shown how the radio window gives a very different view of the Universe from that obtained using light waves only, and in particular how it becomes possible to 'look back' in time so that conflicting cosmological theories can be tested out. Mention is also made of quasars and pulsars which currently pose new problems in astrophysics.

While radio wavelengths enable the Universe to be investigated on the grandest possible scale, X-rays, which have a wavelength far shorter than light waves, may be used to probe the detailed structure of matter on an atomic scale. This aspect is thoroughly discussed in Chapter Five, which surveys the X-ray-diffraction method, showing both how the diffraction patterns are produced and how they may be interpreted to reveal the arrangements of atoms in different types of material.

The last two chapters of the book describe how radio techniques are pushed to the limit in the microwave region of the spectrum and how they gradually merge into quantum techniques. It is here that the interplay of wave and particle aspects of radiation is of great significance, and the discussion of microwave spectroscopy, lasers and masers, gives much insight into the relation between the classical and quantum ideas.

CONTENTS

1   UNDERSTANDING LIGHT   1
    by Dr. A. Hewish

2   LIGHT AND THE ATOM   23
    by Dr. A. Hewish

3   LIGHT FROM THE STARS   37
    by Prof. M. W. Ovenden

4   THE RADIO WINDOW INTO SPACE   56
    by Dr. A. Hewish

5   SEEING THE VERY SMALL   74
    by Dame Kathleen Lonsdale and Dr. Judith Milledge

6   MICROWAVES   103
    by Prof. A. L. Cullen

7   CLOSING THE GAP   125
    by Prof. D. J. E. Ingram

    INDEX   149

ONE

UNDERSTANDING LIGHT

Introduction

Throughout our conscious life, our brains are continually accepting and sifting the flood of signals transmitted by our senses, and it is a fact that the sheer quantity of information received visually far exceeds that obtained by other means. But visible light is merely a fraction of a wide band of similar radiation known as the Electromagnetic Spectrum, which embraces the longest wavelength radio waves at one end and the shortest wavelength γ-rays at the other. One of the triumphs of modern science has been the exploitation of the whole range of this spectrum, so that our 'vision' now extends far beyond the narrow limitations imposed by the human eye. On the largest scale, radio telescopes, for example, are now revealing features of the Universe which are completely hidden from ordinary telescopes, while on a microscopic scale, X-rays disclose the subtle geometrical architecture of atomic structure in crystals with a clarity far beyond the capabilities of conventional microscopes.

In order to appreciate how these extensions to our sight can be achieved, it is necessary to know something about the main features of the electromagnetic spectrum, and the aim of this chapter is to mention some of the relevant physical ideas.

First, then, what is actually meant by the electromagnetic spectrum? Naturally, we are familiar enough with visible light; the processes of evolution have ensured that we are well equipped to sense, and make use of, the radiation which falls most abundantly upon the surface of the earth. But the radiation which reaches the ground is merely a fraction of that arriving at the outer layers of the atmosphere. The colours to which our eyes are sensitive, ranging from red and orange, through yellow, green and indigo to violet, constitute just that fraction of the spectrum which is not absorbed by the atmosphere. That sunlight contains other kinds of radiation is well known to mountaineers and skiers, who rapidly tan at an altitude of a few thousand feet where ultra-violet radiation is considerably stronger than at sea-level. Much higher still, the radiation can become positively damaging due to the presence of X-rays, a type of radiation which is highly penetrating and which can pass right through the human body. How it can do this when it is basically light, but with a 'colour' to which the eye does not respond, will become clearer when the nature of light has been more fully discussed. A more damaging radiation still, which occurs at yet shorter wavelengths, is known as γ-radiation. This radiation is associated with disruption within the atomic nucleus and requires extremely high energy for its release. γ-radiation is produced abundantly deep inside the sun, where it is retained by overlying layers of solar material. It is also generated by hydrogen bombs, which simulate the sun on a smaller scale.

Having considered the extension of the visible spectrum beyond the violet, we should now go in the other direction and enquire what happens when light is 'redder than red'. This is known as the infra-red end of the spectrum. The radiation encountered here is familiar as radiant heat. The glowing element of an electric fire emits far more energy in the form of heat radiation than it does as light. This is readily apparent on switching on an electric fire, when its warmth may be felt long before the element is hot enough to glow visibly. Infra-red rays are much more easily stopped than light rays, a fact which is exploited in the construction of a greenhouse, where use is made of the property that light waves from the sun pass easily through glass and warm the objects upon which they fall. The latter then emit infra-red rays which cannot pass out through the glass again, and thus the temperature within the greenhouse rises.

It should be emphasised at this point that it is important to distinguish between the infra-red radiation and the heat which it produces. Infra-red radiation is simply light of a long wavelength to which the eye is not sensitive. When it is absorbed, for example by the skin, the energy which it carries is transformed into heat. Now visible light, ultra-violet, X-rays, γ-rays, etc., can all be similarly transformed into heat. The reason that we associate infra-red rays with heat is just that a body which is 'red-hot' radiates most strongly in the infra-red portion of the spectrum. It can easily be demonstrated, however, using detectors sensitive to other wavelengths, that any hot body radiates at least some energy over a very wide range of the spectrum.

Proceeding even farther from the red end of the spectrum we come to microwaves, which are radio waves of very short wavelength, and eventually to ordinary radio waves. The human senses are hopeless detectors at this end of the spectrum, which we normally associate with electrical phenomena, although even here the conversion of radiation into heat will be familiar to those who have any experience of radio therapy. For certain therapeutic and industrial purposes it is convenient to apply heat by means of radio waves by the conversion mechanism mentioned in the last paragraph. What makes the process useful is that the heat is generated where the radiation is absorbed most strongly, and this may be deep inside an otherwise inaccessible region. By this means it is possible, for example, to apply heat beneath the skin of the human body.

There is virtually no limit to the electromagnetic spectrum, which stretches, as we have seen, from long radio waves at one extreme, through the visible, to X-rays and γ-radiation at the other end. But we have yet to answer the question—what *is* light? Before attempting any kind of explanation it is, perhaps, as well to digress for a moment and see just what a scientist means when he claims to understand something, for this brings us to the essence of scientific method.

Scientific Understanding

By 'understanding' a phenomenon, a scientist usually implies that he can fit this occurrence into some logical pattern of events which can be summarised in a well-defined physical law. The latter is usually a mathematical relationship between certain quantities. A good example is the Law of Gravitation, which states that two lumps of matter attract each other with a force which depends directly upon the masses, and varies inversely as the square of the distance between them. Application of this simple relationship explains the curved flight of a projectile, the ebb and flow of the tides, and how it is that an artificial satellite can circulate the globe without falling to the surface. All these phenomena could have been predicted from the law of gravitation, and the scientist therefore says that he understands them. But ask him *why* it is that matter attracts other matter, rather than *how*, and no answer is forthcoming. Matter simply behaves in this way, and science can go no farther at the moment.

Now physical laws and concepts are often difficult to grasp, and it is frequently of value to invent some kind of model with which we are already familiar and which also embodies some of the properties expressed in a physical law. The force of gravity, for example, might be imagined as a stretched spring, or piece of elastic, connecting two bodies. We now see why two bodies will fly together unless held apart by another force; or again, by twirling a stone about our head on the end of an elastic string, we begin to appreciate the dynamics of an artificial satellite. Such models, or analogies, are valuable when groping towards an understanding, but they must not be confused with the reality itself. No spring or elastic is infinitely stretchable, as would be required to simulate the law of gravitation more exactly. In the end it is only the laws themselves which count, and our deeper understanding follows when the many implications of a law have been fully worked out and have become part of the fabric of our thought.

It is a mistake to regard a physical law as absolute truth. The law is simply a generalisation, or abstraction, based on the observation of some physical system which sums up the totality of experience so far as it is known. As time goes on and observations become more accurate, the law is seen to be only an approximation to the truth. Some modification is required, and the consequences of the new law may then be worked out and tested experimentally. In this way our knowledge deepens, but it would be a rash man who claimed, ever, to have plumbed reality to its depths.

Scientific theories therefore advance step-wise, and at certain stages in their development it may happen that two quite different concepts appear equally tenable. Being human, the devotees of one particular idea are then prone to view the rival theory with hostility, an unscientific attitude which has frequently impeded the progress of knowledge. As will be seen in what follows, the understanding of light has been reached only through such a painful process of evolution.

The History of Light—Waves Versus Particles

While many properties of light, such as the laws of refraction and reflection, and the action of a burning glass, have been known from antiquity, little record can be found of any serious speculations about the nature of light until the era of the Greeks. It is unfortunate that this talented race, while first-rate thinkers and philosophers, made very poor physicists, owing to their disinclination to undertake experiments. Their general attitude is typified by the outlook of Plato, who imagined the only basic reality to be the mind—the world disclosed by physical observation being merely a somewhat tainted and distorted replica. But even speculation has its value, and Empedocles, of the Platonic school, was near the truth in supposing light to be some kind of emission. His view was rather complicated, and involved an emanation from the eye which then combined with similar emanations from the sun and the object seen. The same notion, but shorn of some needless complexity, reappears some centuries later in the writings of Lucretius, who states:

'It must be a fact that all visible objects emit a perpetual stream and shower of particles that strike upon the eyes and provoke sight. From certain objects there also flows a perpetual stream of odour, as coolness flows from rivers, heat from the sun and from the ocean waves a spray that eats away walls around the seashore. Sounds of every sort are surging incessantly through the air ... so from every object flows a multiform stream of matter, rippling out in all directions.'

It is interesting to notice that two concepts appear to be involved in this passage. In the first sentence light is likened to a hail of shot, or a flight of arrows speeding from the luminous object, while in the last sentence 'rippling out in all directions' brings in the notion of waves such as might be launched by throwing a stone into a pond. Whether or not this is really what Lucretius had in mind, these two concepts have played a central role in the theory of light. At times it has appeared as if one or other must be rejected, but even today a complete understanding of the nature of light contains elements of both ideas.

Passing down the centuries, we find little advance in the basic theory, even after the introduction of optical telescopes early in the seventeenth century. Descartes, something of a scientific predator who was inclined to claim the results of others as his own, imagined light to be a kind of pressure wave transmitted through an elastic substance which filled the entire volume of space. It was only in the mid-seventeenth and early eighteenth centuries that the first break-through came, and this followed a series of careful observations.

That light travels in straight lines is very obvious from the occurrence of shadows. In 1665, however, a book published after the death of the Jesuit father Francesco Maria Grimaldi showed that the formation of shadows is not the whole story. When careful observations are made, it can be seen that a fraction of the light bends round into the shadow region and there gives rise to alternating bands of light and darkness. Grimaldi's original sketches show the effect plainly, and a photograph of the phenomena is shown in Figure 1. The bands are very narrow and a small light source, such as an illuminated pinhole or fine slit, is needed to show the effect. It is therefore not surprising that it remained undetected for so long. The explanation of his observations eluded Grimaldi, although he was near the truth when the bands reminded him of the succession of ripples which occur when a stone is thrown into a pond.

Further information was contained in another book which appeared in 1665, this time by Robert Hooke, just three years after he had left the employment of Robert Boyle to become Curator of the recently estab-

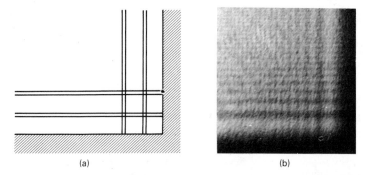

(a) (b)

FIGURE 1 (a) Variations of light intensity at the edge of a shadow as sketched by Grimaldi. (b) Photograph of a similar effect.

lished Royal Society. This book described the colours seen in thin films, such as soap bubbles and oil floating on water. Hooke showed how the colour depended on the thickness of the layer and progressed steadily through the colours of the rainbow as the thickness was increased; but thought that vibrations of the film were involved, the vibrations giving rise to spherical disturbances which somehow combined in different ways to produce colours.

At this time the origin of colour was far from clear, and Isaac Newton appears to have made the first systematic investigations while a Fellow at Trinity College, Cambridge. He purchased a prism at Stourbridge Fair, completely darkened his room save for a small hole in the shutter and allowed the resulting shaft of sunlight to fall upon it. The beam was, of course, spread into the familiar coloured spectrum, and Newton noted how the beam, originally circular in section, was drawn out into an elongated shape, red at one end and violet at the other. He soon concluded that this could be explained only if violet light suffered a greater deviation than red light in passing through the prism. Newton then went on to see if a particular colour would be further changed by its passage through a second prism and wrote:

> 'When any one sort of Rays hath been well parted from those of other kinds, it hath afterwards obstinately retained its colour, notwithstanding my utmost endeavours to change it.'

By means of extremely simple observations Newton thus proved that white light must contain a mixture of all colours, and that colour is not some property imposed upon the light by the medium through which it passes.

This was a great stride forward, although Hooke, a contemporary of Newton, was far from convinced by Newton's careful experiments. During the ensuing debates the nature of light was much discussed, but Newton could do little more than add certain variations to Hooke's concept of vibrations, and in the end he seemed to believe that the particle theory came nearer to the truth. As he speculates in the concluding passages of his 'Opticks' we find him asking, 'Are not the rays of light very small bodies emitted from shining substances?' This is, perhaps, more surprising when we remember that Newton had also observed Grimaldi's diffraction in the V-shaped opening between the blades of a pair of crossed knives, but he attributed this phenomenon to a force exerted on the particles as they sped close by the material of the blades.

One of the reasons why Newton found it difficult to believe in waves was the discovery of *polarisation* by Christian Huygens. In order to see what is involved here it is necessary to appreciate that the propagation of a disturbance in the form of a wave is generally one of two kinds. One type, a compression wave, is best visualised in terms of a coiled spring as shown in Figure 2 (a). We suppose that the spring is supported horizontally and one end of it is suddenly pushed in, and then released. When this happens a wave disturbance is launched along the spring as illustrated. It will be noted that the displacement of individual turns of the spring takes place in the direction along which the wave is travelling, and this is called a *longitudinal* wave. Air is compressed longitudinally when a sound wave passes through it, and this kind of disturbance was familiar to Newton and his contemporaries. Now when we launch a wave along a string, as shown in Figure 2 (b), the string is displaced at right angles to the direction of travel. Whether the motion is up and down or horizontal depends on how the string is initially displaced, and the precise direction is maintained thereafter. This kind of wave is called a *transverse* wave, and the particular direction which specifies the displacement of the string is known as its sense of *polarisation*. A transverse wave can thus be horizontally or vertically polarised, or some mixture of the two corresponding to oscillation in an intermediate direction. When a wave is polarised transversely it is possible to find obstacles which will let a wave polarised in a particular direction pass through while impeding the passage for a different polarisation. If a string runs through a vertical slit a vertically polarised wave is allowed through while a horizontally polarised wave is completely stopped, as shown in Figure 2 (c).

The fact that the vibrations of a light wave are tranversely polarised is easily demonstrated using two sheets of polaroid as illustrated in Figure 3. Natural light from the sun or an electric bulb contains a mixture of all polarisations, but, by passing it through a polaroid, all except one of these are extinguished. If the polarised light now traverses a second polaroid

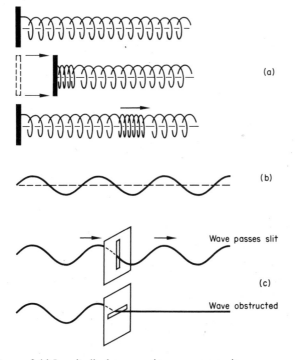

(a)

(b)

Wave passes slit

(c)

Wave obstructed

FIGURE 2 (a) Longitudinal compression wave on a spring.
(b) Transverse wave on a string.
(c) Effect of a slit on a vertically polarised transverse wave.

arranged to be at right angles to the first the light will be completely extinguished. Effects similar to these were known to Huygens, since they also occur when light passes through certain crystals, such as calcite or Iceland spar. Unfortunately Newton was familiar only with the longitudinal compression waves of sound, and while he was close to the truth when he conjectured that rays of light needed 'sides' to account for Huygens results, the transversality which this implied led him only to reject more vigorously the wave ideas of Huygens.

Huygens was, himself, an advocate of the wave theory of light, and he showed a remarkable flash of genius in enunciating the principle that any point on a luminous disturbance may be regarded as the source of a new wavelet which propagates from it spherically. By adding up these secondary wavelets, it is possible to work out what happens when a wave disturbance meets some obstacle in its path. Using this principle, it is a simple matter to see just how the well-known laws of reflection and refraction come into being. The case of reflection is made clear in Figure 4, where a light wave striking a mirror is imagined to set up a succession of spherical wavelets at each point where the incoming wave hits the reflecting surface. The first wavelets produced grow in size during the time taken for the whole wavefront to reach the mirror, and these disturbances combine to give the reflected wave. It was a pity that no one saw how well Huygens' postulate could also account for the more complicated details of the diffraction of light into a shadow. Newton's authority was such, however, that his advocacy of the corpuscular theory swayed scientific opinion from the wave theory for nearly a century.

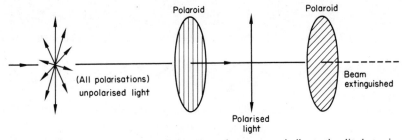

FIGURE 3 Light is a transverse wave. Polaroid acts in a manner similar to the slit shown in Figure 2 (c).

The next advance came in 1801, and was contained in a paper to the Royal Society by the gifted Englishman Dr. Thomas Young. Young was not skilled in casting his results into a mathematical form, and for this reason his work tended to be overlooked, yet his observations laid the foundations for a correct theory. In particular, he showed that under certain conditions light can actually cancel itself out. That is, by adding light to light we can create darkness. Even now this may seem a strange result, and it is not difficult to see why Young had his critics. As often happens, however, the best scientific experiments are the simplest, and one of Young's observations is certainly a good example. The arrangement is shown in Figure 5, which illustrates how the light from one pinhole illuminated two others behind which a screen was situated. With this arrangement, the light was found to fall on to the screen in a series of bands with darkness in between. With one of the final pinholes covered, however, the light was evenly spread over the screen. This observation was explained by Young in his Principle of Interference, which stated that the crest of a light wave from one of the pinholes can exactly cancel the trough of a wave from the other pinhole, giving no illumination. It is easy to see

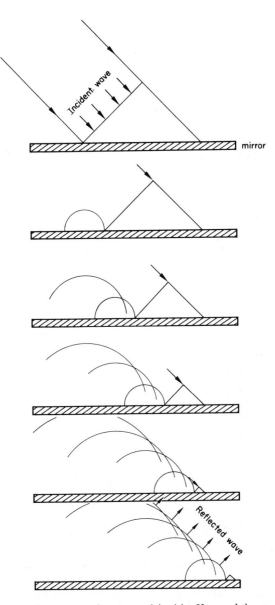

FIGURE 4 Reflection of light at a mirror as explained by Huygens' theory of secondary wavelets.

how this comes about from the sketch in Figure 5. The light waves from the first pinhole generate, at the pair of pinholes, twin secondary waves whose crests and troughs add together or cancel each other out according to their position on the screen and the relative distance of the point under consideration from each of the two pinholes. So long as light is considered as waves, the explanation is simple and convincing.

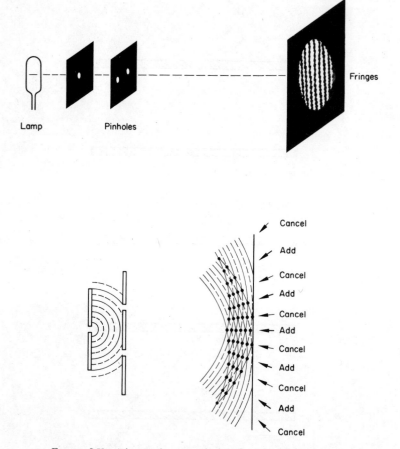

FIGURE 5 Young's experiment on the interference of light waves.

The particle theory of light received a further blow some years later at the Paris Academy when the topic of diffraction was proposed as the subject of a prize essay in 1818. The prize was won by Jean Fresnel, who combined the ideas of Huygens and Young in a mathematical form and showed how this could account readily for all the known phenomena of diffraction. The Academy was shocked to find the particle theory so thoroughly defeated. Doubts were raised by the eminent mathematician

Poisson, who deduced, from Fresnel's theory, that light should be found at the exact centre of the shadow cast by a circular object. This naturally seemed to be absurd, but the prediction was beautifully verified by Francois Arago, and the episode merely added further laurels to Fresnel's cap, while hammering another nail into the coffin of the corpuscular theory. The reason for a bright spot at the middle of a circular shadow is that light is diffracted into the shadow region and, at a point of symmetry in the exact centre, all the waves are exactly in step and combine to give a bright spot. Using Fresnel's theory, it can be estimated that the illumination at this point is only marginally less than if the shadow-casting obstacle were completely removed!

Fresnel and Arago made other experiments into Young's phenomenon of interference, and for some of these they used light which had been polarised by passing it through calcite. They found that rays polarised in different directions at right angles could never be made to interfere, and from this Young correctly deduced that the vibrations of light waves must be transverse, rather than longitudinal, as had been assumed hitherto.

To show that the behaviour of light can be explained in terms of waves is a major stride forward, but there is still the question—waves in what? All the waves with which we are familiar—ocean waves, sound waves, waves on a string, etc.—require some medium in which to travel, and the wave itself represents some displacement of matter. But light will travel across empty space, where no material exists to vibrate. Faced with a problem such as this, the physicist invokes some model to help him along, and at that time the whole of space was assumed to be filled with 'aether', a kind of colourless, weightless jelly which, however, was able to vibrate and transmit wave motion.

The general notion of an aether had been present since antiquity, but in the nineteenth century some of the difficulties of this concept became apparent. One obvious objection to the presence of an aether was the fact that planets must be rushing through it at immense speed, yet with no resistance to their motion. One possible idea here was to suppose that the vibrations of the light waves corresponded to velocities much greater than those of the planets, and that the aether might have elastic properties similar to pitch—the latter is plastic and will flow like a liquid if given a sufficiently long time, but for rapidly applied distortions it behaves like an elastic solid. However, as we shall see later, there is no need to invoke an aether with complicated properties in order to construct a perfectly satisfactory physical theory.

The Speed of Light

No mention has yet been made of the speed at which light disturbances travel from one point to another. To a good approximation, light appears to travel instantaneously. When we count the seconds between seeing a lightning flash and hearing the thunderclap it is obvious that sound takes a definite time to traverse a certain distance; but what about the light? When we see the flash, are we witnessing an event which actually took place some fraction of a second earlier? The first attempt to make a direct measurement of the velocity of light appears to have been undertaken by Galileo. He stationed two men with lanterns a few miles apart and arranged that the second man should uncover his light when he saw a flash from the first man's lantern. In this way Galileo hoped to be able to detect a time interval between the first man uncovering his lantern and then seeing the second man's light. But the speed of light was much too great to be detected in this way.

The first real evidence about the speed of light was obtained, almost accidentally by the Danish astronomer Olavs Römer in 1676 from careful measurements on the eclipses of Jupiter's satellites. As each satellite revolves about the planet, it appears and vanishes at regular intervals, which can be measured with high precision, and hence the occurrence of future eclipses can be accurately predicted. Römer noticed that there were definite errors between the predicted eclipse times and those which were actually observed, and he put this down to the varying distance between Jupiter and the earth. When Jupiter was receding from the earth the eclipses were progressively later than they should have been, while they were correspondingly earlier when Jupiter approached the earth. Römer was able to account for the errors if light travelled through space at about $3 \cdot 1 \times 10^8$ metres per second. At this speed it is not surprising that Galileo's lantern men should have failed to detect the few millionths of a second travel time of the light between them!

A remarkable confirmation of this estimate came about fifty years later when Bradley attempted to ascertain stellar distances by measuring the shift in position of a star when observed from opposite sides of the earth's orbit around the sun. Bradley's results made sense only if the small position changes which he found were not caused by parallax, which he expected, but by an effect known as aberration which comes from combining the velocity of light with that of the earth in its orbit.

Really accurate measurements came later in the nineteenth century when Fizeau used a toothed wheel rotating at great speed to chop a light beam into narrow pulses which travelled a certain distance before being reflected back to the wheel again. On their return the wheel had advanced a small amount, and the rotation speed could be adjusted so that the reflected pulses struck a tooth rather than the gap between adjacent teeth. The reflected beam was then cut off, a clearly visible effect, and the speed of light could be calculated. Modern methods are essentially variants of the same idea and give a value of almost exactly 3×10^8 metres per second; Römer's result, although not really believed for fifty years, was a remarkably good attempt.

Light and Electricity

Perhaps the greatest breakthrough in understanding the nature of light came from quite a different direction. It sometimes happens in scientific investigations that a discovery in one field can be immediately related to another phenomenon with which there has been no obvious connection, and so it was with the theory of light. Michael Faraday devoted most of his life to a study of electric and magnetic phenomena. He was a skilled experimenter and possessed great physical insight, but he was no mathematician. To account for the forces which he knew to exist between magnets and between electrically charged bodies, he imagined space to be filled with *lines of force.* Just what this implies is best seen by considering a simple example. Suppose we have a single charged particle, such as an electron. If we now imagine the electron to be at the centre of a set of lines

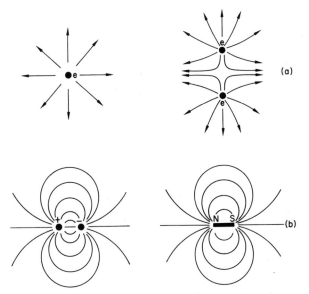

FIGURE 6 (a) Lines of electric force near an electron, and an electron pair.
(b) Lines of force around an electric dipole and around a magnet.

as illustrated in Figure 6 (a), then another charged particle near it will experience a force in a direction indicated by the lines, and of magnitude proportional to the density of lines in space at that point. The clustering of the lines near the electron shows how the force increases the closer we get to the electron. Lines of force need not be straight; for equal negative and positive charges, such as an electron and a proton, a curved pattern is obtained, as shown in Figure 6 (b) and the lines of force around a bar magnet are much the same.

When first considering electric and magnetic forces, Faraday seemed to think that the presence of an electric charge caused the aether to be strained in some way, as if space were filled with a stretched elastic jelly. He later modified this idea to one of tensions in the lines of force, and it was then only a short step to the intuitive guess that the wave vibrations of light could be thought of as vibrations of the lines of force. Lecturing in 1846, Faraday said:

'The consideration of matter under this view gradually led me to look at the lines of force as being perhaps the seat of the vibrations of radiant phenomena. . . . The view which I am so bold as to put forth considers, then, radiation as a high species of vibration in the lines of force. . . .'

Faraday's ideas were taken up by Clerk Maxwell, an able mathematician, who, in 1856, published a paper 'On Faraday's Lines of Force'. Maxwell put together the known results of electricity and magnetism and expressed them in an elegant set of equations, a mathematical structure

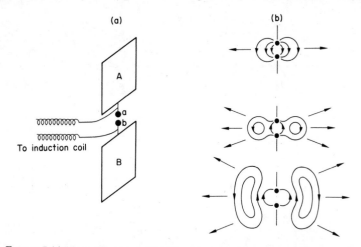

FIGURE 7 (a) The apparatus used by Hertz for generating electromagnetic waves. (b) Lines of force at successive instants near an oscillating dipole.

which is regarded as one of the finest achievements of physics. One of the predictions of Maxwell's theory was that any disturbance of a set of lines of force would launch a wave in just the same way as a wave can be propagated along a stretched string. Furthermore, the velocity of the disturbance predicted by the theory had exactly the same value as the experimentally measured velocity of light. In other words, light waves can be identified with fluctuating electric and magnetic forces, and this is the best answer that physics can give regarding the nature of light.

It can be imagined that some time elapsed before these complicated notions were generally accepted. The demonstration, by Heinrich Hertz in 1887, that real electrical waves, having all the properties calculated by Maxwell, could be generated, finally convinced the sceptics, particularly those on the Continent. The main features of the apparatus used by Hertz are shown in Figure 7 (a). The flat metal plates A and B were attached to an induction coil which caused sparking to take place in the gap between the spheres *a* and *b*. When each spark occurred there was a rapid oscillation of charge between *a* and *b* so that the spheres were alternately positively and negatively charged, the time taken for a change of sign being about one-hundredth of a millionth of a second. Now we can calculate, by Maxwell's theory, just how the lines of force will behave in this situation. At some instant the opposite charges on *a* and *b* will give a pattern of lines of force similar to those shown in Figure 6 (b). But, as the charges alternate, so the lines of force must change, and the sequence will be roughly as illustrated in Figure 7 (b). Bundles of lines apparently emerge and break off from *a* and *b*, giving a succession of closed loops which travel outwards at the velocity of light. Hertz demonstrated that these bundles existed by setting up a ring of wire at the other side of the laboratory and obtaining discharges across a small spark-gap. This was the first radio transmitter in the world. At a sufficiently large distance from the

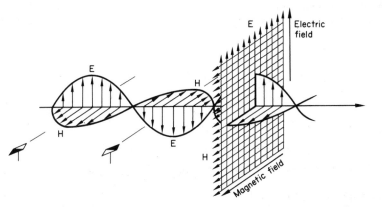

FIGURE 8 Electric and magnetic lines of force associated with an electromagnetic wave.

generator the spherically diverging waves have spread to such an extent that their curvature can be ignored, and they are then said to be plane waves—that is, the fluctuating electric and magnetic lines of force are the same, instantaneously, at all points across a flat surface perpendicular to the direction of travel. One of the easiest forms of wave motion to think about is the simple harmonic wave, in which the vibrations vary regularly along the path of the wave. Many natural waves, such as ocean waves or ripples or a pure note in sound, approximate to this ideal. The lines of electric (E) and magnetic (H) force for a simple harmonic electromagnetic wave are sketched in Figure 8, and this is the best physical picture which can be given for a light wave. The idea of electric and magnetic forces drifting through space is difficult to grasp, but it is not so hard to imagine the periodic fluctuations of a magnetic compass needle which would occur if a sufficiently sensitive compass were placed in the passage of such a wave.

Hertz was able to show that his disturbances behaved in much the same way as light. He could refract the waves by passing them through a large prism made of pitch, reflect them from flat surfaces, and he also showed that they were polarised. The latter was done by erecting a screen of parallel wires and demonstrating that the waves easily passed through it when the wires were vertical (at right angles to the line joining the spheres *a, b*) but not when the wires in the screen were horizontal. This is an exact analogy to the case of waves on a string, as in Figure 2 (c).

Some Properties of Waves

So far nothing has been said about the actual *wavelength* of electro-magnetic waves, which is defined as the distance between successive crests. Since the waves are moving along at some definite velocity, successive wave crests will pass a stationary point at definite intervals of time, and the number which pass in one second is known as the *frequency* of the waves. A cork floating in water, for example, would bob up and down at the wave frequency. Now it is self-evident that the speed at which the waves are

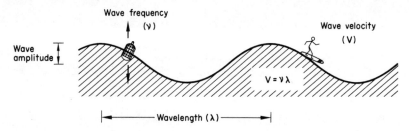

FIGURE 9 Relation between wavelength, frequency and velocity of a sinusoidal wave.

travelling is simply the wavelength multiplied by the frequency, so that $V = \nu\lambda$, where V is the velocity, ν the frequency and λ the wavelength. These relationships are shown in Figure 9.

One of the main conclusions of Maxwell's theory is that the velocity of electromagnetic waves through empty space is always the same, and is in no way affected by the wavelength. It therefore follows immediately from the simple mathematical relation $V = \nu\lambda$ that electromagnetic waves of small frequency have a long wavelength, while waves of high frequency have a short wavelength. Now it is just this property which differentiates radiation over the spectrum and causes it to behave as X-rays, light waves or radio waves. The varying nature of radiation is simply a matter of its wavelength. Starting with radio waves, the 'medium' wavelengths are from 200 to 600 metres, corresponding to frequencies of $1 \cdot 5 \times 10^6$ to 5×10^5 crests or *cycles* per second. (Recently it has been decided to refer to frequencies in terms of hertz so that a frequency of 50 cycles per second (50 c/s) is now called 50 Hz.) V.H.F. (very high frequency) and Television use much higher frequencies of around 100 million hertz corresponding to wavelengths of a few metres, while radar uses 'microwaves' only a few centimetres in length. For reasons which will be discussed in later chapters, it becomes difficult to handle wavelengths much shorter than this by radio techniques, and so we pass out of the radio spectrum into the infra-red.

For light waves the frequency is even higher, and the wavelength has become so small that half a million waves can be packed into one centimetre. And so it goes on right down the spectrum to gamma rays, which have the shortest wavelength of all. A summary of the spectrum in terms of wavelength and frequency is shown in the table, and it is a resounding achievement of Maxwell's theory that all these radiations can be described in a similar way.

THE ELECTROMAGNETIC SPECTRUM

Name	Wavelength in metres	Frequency in hertz
Radio	Greater than 1	Less than $3 . 10^8$
Microwave	$\frac{1}{1000}$ to 1	$3 . 10^{12}$ to $3 . 10^8$
Infra-red	10^{-6} to 10^{-4}	$3 . 10^{14}$ to $3 . 10^{12}$
Visible	$5 . 10^{-7}$ to 10^{-6}	$6 . 10^{14}$ to $3 . 10^{14}$
Ultra-violet	10^{-8} to $5 . 10^{-7}$	$3 . 10^{16}$ to $6 . 10^{14}$
X-rays	10^{-11} to $1 . 10^{-8}$	$3 . 10^{19}$ to $3 . 10^{16}$
γ-rays	10^{-14} to 10^{-11}	$3 . 10^{22}$ to $3 . 10^{19}$

A famous scientist once remarked that 'physics is bottomless', implying that we can never really know the whole truth about nature. However, the theoretical structure erected by Maxwell seemed flawless and invincible until the turn of the century when some awkward phenomena began to be discovered which did not fit into the wave picture at all. The result was a major upheaval in physical concepts which brought in its train, as always, a deeper understanding of reality. This chapter will conclude with a discussion of modern ideas which, while not refuting Maxwell's ideas, have shown that there is another facet of the truth which cannot be ignored and may often be of dominating importance. Before embarking on this, however, it will be profitable to describe a few more properties of waves which will be mentioned in other sections of the book.

Every schoolboy knows that 'light travels in straight lines', but this turns out to be only an approximation, as Grimaldi found when he noticed that shadows were not perfectly sharp (see Figure 1). The bending of light into a shadow is known as *diffraction*, and this behaviour is of extreme importance when it is desired to generate a beam of radio waves for some purpose, such as radar or communications. In order to understand diffraction it is convenient to recall Young's experiment on interference, in which he showed how two light beams could cancel each other out providing that the wave crests of one beam arrived exactly on top of the trough of the other (see Figure 5).

Suppose we allow light waves of some particular wavelength to illuminate a hole in an opaque screen, as illustrated in Figure 10. It will be remembered that Huygens' postulate, mentioned earlier on page 7, states that any point on a wavefront can be regarded as a new source of waves. A detailed mathematical proof of this idea was eventually carried out by Kirchhoff, and this way of dealing with a wave disturbance is extremely valuable in diffraction problems. In our case we can imagine the light wave as it arrives at the hole in the screen to be divided up into a large number of small sources, each one of which is radiating its own set of *spherical* waves. To find where the light goes on the far side of the screen, it is only necessary to add up the (hypothetical) secondary waves. Now if we consider waves going straight ahead, in the direction of the original beam, it is

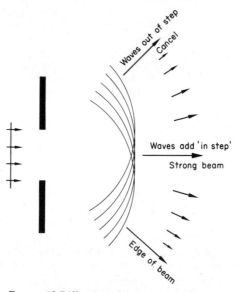

FIGURE 10 Diffraction of light by a circular hole.

clear from Figure 10 that, at a fairly large distance from the screen, all the secondary waves have travelled approximately the same distance. Since these waves all started out in step, this means that they also arrive in step, and consequently add up to produce a large intensity. At some angle to the screen, however, we get a very different result. Again from the diagram it is obvious that secondary waves from the top of the opening travel a shorter distance to a given point above the axis of the beam than do those from the bottom. When adding up the secondary waves we must now make allowance for this progressive difference in path length, the magnitude of which depends upon the width of the opening in the screen. Clearly the light intensity must fall, as the angle from the central beam increases, since the secondary waves are tending more and more to cancel each other as they arrive out of step. It is not difficult to show, mathematically, that the beam is virtually extinguished at an angle such that the difference in path length between secondary waves starting from the extreme edges of the hole is about one wavelength. That is, the smaller the opening, or the longer the wavelength, the greater will be the spreading of the beam. The light does not cut off smoothly at the edge of the beam but, as both calculation and observations show, there is a small fluctuation of intensity producing 'fringes' which surround the main beam, and it was an effect like this which Grimaldi saw.

The fact that light and radio waves do not travel in exactly straight lines is of considerable importance in practice. One example, which is closely analogous to the case discussed above, is what happens in an astronomical

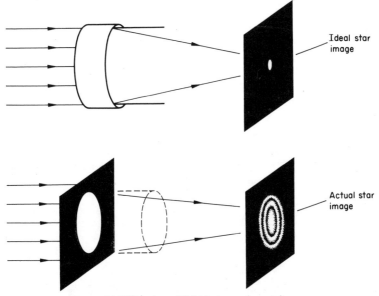

FIGURE 11 Diffraction of light in traversing a telescope.

telescope. Ideally, the parallel rays from a distant star would be focused to a spot of negligible size, as shown in Figure 11. But the lens at the front of the telescope can *itself* be regarded as a circular hole through which the incoming waves must pass. Now we have just discussed how the light spreads out when it passes through a hole, and the real situation in the telescope is as sketched in the diagram. The point focus is smeared out into a disc, and hence we get a blurred picture of the star. This effect, which cannot be avoided, is so serious that it is not possible to measure the sizes of stars directly. The blurring caused by diffraction can only be reduced by increasing the diameter of the telescope, and there are obvious limitations imposed by financial and engineering difficulties. Exactly the same thing happens in radio telescopes, but the effect is much worse, since the wavelength is so much longer.

Very interesting effects take place when a plane light wave falls upon a regular arrangement of many holes in an opaque screen. Now we have to combine both the beam-spreading effect, which has just been discussed, and the superposition of many beams from the different holes. If each hole in the arrangement is sufficiently small, then the beam from each opening will be spread over a considerable range of angle and, at a large distance from the array, the beams will overlap. Now in certain directions the waves in the different beams will be exactly in step. This happens when the path length of the light from adjacent holes differs by a whole number of wavelengths and, of course, the result is a strong beam in this direction. The general situation is sketched in Figure 12, which shows how the array

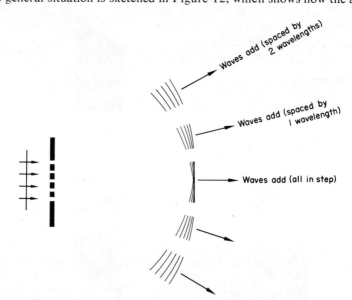

FIGURE 12 Formation of a series of beams by a diffraction grating.

generates a family of separate beams. One type of regular array of apertures in a screen acting in this way is known as a *diffraction grating*, and the device affords an excellent method of analysing a mixture of wavelengths. Each wavelength gives its own set of beams and, by measuring the angle of the beams, the wavelengths can be calculated.

The same principle is used in X-ray-diffraction studies of the arrangement of atoms in a crystal. Each atom diffracts waves in much the same way as the holes we have just considered, and the resulting pattern of beams can be used to find the spacings of the atoms in the crystal.

A further wave phenomenon of considerable importance is the Doppler effect, in which we are concerned with waves emitted or received by moving bodies. When the waves are travelling in some medium, like water waves, or sound waves in air, the Doppler effect is quite easy to understand. For electromagnetic waves, however, there *is* no obvious medium, and the situation is more subtle. More will be said about this in the next chapter, but it is worth stating here that studies of the Doppler effect have been crucial in clarifying the mystery of the aether.

Returning to waves in water, Figure 13 (a) shows an arrangement which emits waves of a certain wavelength in all directions. A cork floating on the surface will just bob up and down at the wave frequency as the waves travel past. However, an observer in a boat travelling *towards* the source of waves will encounter more waves in a given time, and for him the frequency is apparently *increased*. Conversely, the frequency would apparently decrease if the boat travelled away from the source. Clearly, the distance between the wave crests is not altered in any way, but the apparent wave velocity now depends upon the speed of the boat.

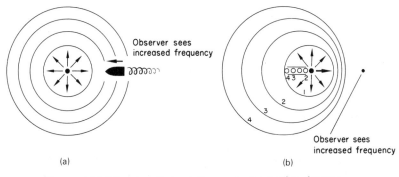

(a)   (b)

FIGURE 13 (a) Doppler effect—stationary source, moving observer.
(b) Doppler effect—moving source, stationary observer.

A slightly different situation arises if the observer is stationary and the wave emitter is in motion. Referring to Figure 13 (b), we see that the emitter is always travelling towards or away from the waves it emitted earlier. Now there is a real change of wavelength but the velocity of the waves is unchanged. The frequency of the waves is increased or decreased as before, but the overall picture is not the same as when the observer was in motion. This form of the Doppler effect is very familiar, as the sudden drop in pitch (frequency) which can be heard from the platform when a train, sounding its horn continuously, rushes through a station.

The Doppler effect finds many practical applications. One of the best known nowadays is the radar speed trap, in which radio waves are bounced off a moving vehicle in order to measure its speed. Before leaving this topic it is important to point out that the difference between the wave pictures due to consideration of the motion of the observer or of the emitter depends upon the fact that the waves themselves always travel at the same speed *in the medium*. Consequently, the Doppler effect for electromagnetic waves can be called in to give evidence about the mysterious aether, and we shall see in the next chapter how this led to a startling revolution in scientific thought. But, strangely enough, this was not the only upheaval to be initiated by experiments using light waves.

Around the turn of the century some awkward, but very straightforward experiments began to cast doubt on the wave theory itself. Prominent among these was the photo-electric effect. When light of sufficiently short wavelength falls on a metal surface electrons are liberated and, if the experiment is performed in a vacuum, these can be collected and measured. Now the energy required to tear an electron from the atoms of a metal can be estimated, and it is also simple to calculate the energy in a light beam shining on the metal. When this is done for a weak light source, the result is that we expect no electrons to be liberated at all! The only way round the difficulty is to suppose that the beam can somehow concentrate its energy at particular points, thus knocking out an electron here and there, rather than falling uniformly over the whole surface. Going right back to Newton's corpuscular theory, we might suppose that the beam is something akin to a hail of shot, in which each particle has enough energy to knock out a single electron.

This strange return to earlier ideas formed another major upheaval in physical concepts. To go further into this topic here would take us rather beyond the scope of the present chapter, but more discussion will be given later on when we look more closely at the relation between atoms and radiation. However, it may be as well to point out now that there is no simple answer. We have to get used to the fact that light can behave both as a wave and as a particle. Neither concept is sufficient by itself, and this demonstrates very clearly how inadequate physical models can sometimes be. The reality we call electromagnetic radiation is more subtle than any single model which we can yet invent. Perhaps what we regard as reasonable, i.e. intuitively satisfactory, is just a question of familiarity, and later generations may not find any difficulty with the dual concept.

TWO

LIGHT AND THE ATOM

The Ultra-violet Catastrophe

In Chapter One we saw how experiments on diffraction—the bending of light round the edges of some shadowing obstacle—showed clearly that light disturbances travel through space in the form of waves. This concept of the wave nature of light thus replaced the conflicting notion that a light beam could be likened to a stream of particles. But it was also mentioned how, long after the establishment of the wave theory, some rather strange phenomena connected with the release of electrons from the surface of a metal when illuminated at sufficiently short wavelength showed that the wave theory did not express the whole truth about light. The resolution of this difficulty takes us into the heart of modern ideas about matter, and in this chapter we discuss some of the main features of the interaction of radiation with different materials.

Before starting, however, it is worth mentioning a further difficulty encountered by the wave theory, a difficulty which was known by the diverting title of 'the ultra-violet catastrophe'. In Chapter One it was explained how heat radiation consists of electromagnetic waves covering a very wide range of wavelength. Red hot metal, for example, glows brightly and obviously gives out plenty of light. But it also emits a great deal of heat at infra-red wavelengths, and sensitive radio receivers show that much longer radio wavelengths still are emitted. Much study has been devoted to this question of heat and light. The general relationships between heat and energy, known as thermodynamics, were thoroughly worked out during the nineteenth century. One aspect of these investigations was the consideration of the statistics of countless billions of atoms in violent motion—the physical model appropriate to a gas. In a case like this it is quite impossible to calculate what happens by following the tracks of individual atoms, since the numbers are far too large to handle. The best that can be done is to assume that each particle moves according to the well-established laws of dynamics, just like a billiard ball, and then to see if the *average* behaviour of a typical atom can be deduced. When this has been done, it is then possible to work out how the bulk properties of the gas, for example, the pressure, will change when heat is added to the system. This approach, known as Statistical Mechanics, proved to be most successful in explaining the behaviour of large collections of atoms, and the theory could also be applied to heat radiation. The arguments are too involved to be summarised here, but since bodies composed of many atoms can be heated by shining radiant energy on to them, one might expect that there is also some connection between the heat radiated *by* a body and the motion of its constituent atoms. This is indeed the case, and the theory as developed by Rayleigh and Jeans resulted in the famous Rayleigh–Jeans law shown in Figure 14. The energy density of the heat radiated (that is the energy contained in a given band of wavelength or frequency) should increase indefinitely as the wavelength becomes shorter. Now it is very easy to measure the energy content of radiation at different wavelengths, and the Rayleigh–Jeans law is obeyed very exactly at wavelengths in the radio band. But serious deviations from the law are found at optical wavelengths where the energy density begins to fall rather sharply from the predicted values. This is just as well, since, according to theory, a red-hot

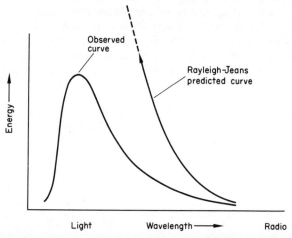

FIGURE 14 Variation of energy with wavelength in heat radiation.

poker would be a strong emitter of hazardous X-rays and γ-rays. Because the theory predicted ridiculously high energies in the ultra-violet and shorter-wavelength regions of the electromagnetic spectrum, this failure of statistical mechanics was called the 'ultra-violet catastrophe'.

The Quantum Theory

Obviously there was some mistake in radiation theory at short wave-lengths, and a way round the difficulty was first put forward in 1900 by Max Planck of Berlin University. Planck suggested that radiation itself possessed some kind of 'atomicity'; all types of energy, in his view, came in units called 'quanta' which could not be subdivided. This notion goes right against the grain at first sight, since, on a human scale, all types of energy seem to be entirely continuous. When we slide a small object on a smooth surface it seems ridiculous to suggest that there is any kind of limit to how *slowly* it can be moved. Yet this is what Planck proposed. If the object is stationary, then it has no energy of motion at all. Now if we give it the least possible energy, just one quantum, then it will start moving at some definite speed and, on the new theory, any speed less than this just cannot happen; 'allowed' speeds are all greater than this basic minimum speed. Going slightly beyond Planck, for the moment, we can calculate this minimum speed if we imagine the object confined to move on a table 3 feet wide—the reason for stipulating the width of the table will become clearer later on. In this case, then, the least possible speed is about one million million million million millionth part of one mile per hour. So ridiculously small, in fact, that any noticeable motion is such a large multiple of the minimum speed that *any* speed appears possible. If we could slide a single electron, however, the least speed would be about one-hundredth of a mile per hour which, while still small, is very much larger.

We shall return to these strange 'quantum' ideas presently. Now, since radiation is also a form of energy, it, too, must be 'quantised'. The least amount of radiant energy which can be absorbed or emitted by any body is again one quantum. These units of radiant energy are called 'photons' and, according to Planck, electromagnetic waves can only have energies which amount to at least one or more photons. When statistical mechanics was worked out again adopting the photon concept, rather than assuming energy to be continuous, the predicted radiation density was found to be in exact agreement with observation, and so Planck's quantum theory was firmly established.

One very important fact about the photons associated with a wave is that their energy is directly proportional to the frequency. This is expressed in the famous equation $E = h\nu$, where h is Planck's constant and ν the frequency. Thus at radio wavelengths where the frequency is relatively small the photons have small energy, but their energy increases steadily all the way up the frequency scale to γ-rays, where the photons have a much larger energy. This means that the atomicity, or particle nature of radiation, should be more apparent at short wavelengths, a feature which is of importance in explaining the photo-electric effect mentioned at the end of Chapter One. It is an experimental fact that, when ultra-violet light falls on a metal, electrons are liberated, but this does not happen for visible light. The explanation is that one photon of ultra-violet light has sufficient energy to tear one electron from the metal, whereas one photon of visible light, of smaller frequency, has not.

Waves and Particles

The idea of photons is quite alien to the wave theory which accounted so successfully for the phenomena of interference and diffraction. On the wave theory the energy is spread out over the whole wavefront and cannot possibly localise itself as one photon. So it seems as if the quantum idea is a return, once more, to Newton's corpuscular theory of light. Do the photons, after all, travel through space like the 'very small bodies emitted from shining substances' envisaged by Newton? The answer to this question is rather subtle, but the blunt truth is that light is, in a sense, composed of both waves *and* particles. Since light exhibits some properties which clearly belong to waves, and others which equally clearly belong to particles, then the physical reality we call light, or more generally electromagnetic radiation, must be some mixture of both.

The trouble is that we are not at all familiar with realities of this sort. A rather poor analogy, which may nevertheless be helpful, is provided by the substance known as 'bouncing putty' which has been sold as a novelty for some years. This material, when kneaded in the hand, seems just like putty or plasticine; it can be moulded into any shape with no trouble. But roll it into a ball and drop it on to a hard surface, and it bounces as though made of rubber. The material is therefore either plastic, and permanently deformable, or elastic, according to how we treat it. The key in this case is just one of time-scale. On a long time-scale, when deformed slowly, the substance is plastic; on a short time-scale sufficiently rapid deformations cannot be sustained and the elastic properties are dominant. But there is no answer to the straightforward question—is it elastic or plastic? It is clearly either, depending on how we treat it.

A new slant to the difficulty of the wave-particle concept of light was introduced in 1925 by de Broglie. Arguing from analogy, often a most fruitful exercise in theoretical physics, de Broglie suggested that if light waves sometimes behave like particles, then it is possible that particles might also be made to exhibit wave properties. From a discussion involving Einstein's equation connecting mass and energy, which came from the theory of relativity, it seemed likely that very light particles should show wave effects most easily. Nowadays there is a long list of the so-called 'fundamental particles' which result from smashing atoms and nuclei into the smallest possible pieces. The electron, the smallest unit of electric charge, was the first such particle to be discovered, and it was a natural choice as a candidate which might give some evidence of behaving as a wave. As it happened, the crucial experiment had already been made, and de Broglie's hypothesis immediately accounted for a strange result obtained when electrons were reflected by a metal surface. When an ordinary ball is thrown on to a flat surface it rebounds so that its angle of flight is exactly the same, relative to the surface, as that at which it arrived (see Figure 15); but when the same experiment is carried out using electrons it is found that more electrons leave the surface in certain directions than in others—there seemed to be *preferred* directions for the reflected electrons. Now this cannot be explained if the electrons behave like ordinary balls or particles. The results are accounted for, however, if we suppose that the electrons are *diffracted* by the metal surface which consists of metal atoms set out in a regular crystalline grid. Exactly the same behaviour is exhibited when light waves are split into several beams by a diffraction grating (see Chapter 1) or when X-rays are diffracted by crystals, the subject of another chapter in this book. Because the spacing of the atoms in the metal can be calculated from other information, it is possible to use the electron-diffraction results to estimate the 'wavelength' of electrons, and it turned out that de Broglie was entirely correct. The wavelength λ was given by $\lambda = h/p$, when h is Planck's constant and p is the momentum of the electron. So the electron, at any rate, can show diffraction, a typical wave property.

Now it might be argued that an electron is so small that it scarcely counts as a real particle. It also carries a negative electric charge which makes it somewhat different from a billiard ball. Doubts of this kind led to a repetition of the same experiment using molecules of hydrogen and helium. Here we really are dealing with ordinary matter—the molecules being uncharged and a few thousand times more massive than the electron. Once more it was found that the molecules bounced off a crystal surface in preferred directions according to the rules of diffraction, the 'wavelength'

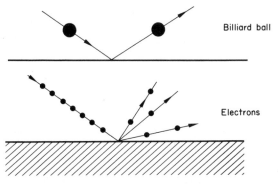

Billiard ball

Electrons

FIGURE 15 Comparison of the behaviour of billiard balls and electrons after collision with a reflecting boundary.

again being given by de Broglie's formula. So the conclusion was quite inescapable—particles of matter behave as waves under certain conditions, just as light waves must sometimes be regarded as particles. In other words, when considering what matter actually *is*, both its wave nature and particle nature are important—there is no such entity as a pure wave, or a pure particle, although, by going to suitable extremes, good approximations can be made. A billiard ball, for example, is so heavy that the wavelength, using de Broglie's formula, is immeasurably small for any sensible velocity. We simply cannot make one travel slowly enough for any wave effects to be apparent. On the other hand, the photons associated with radio waves are so tiny that any detectable wave involves countless photons and it is impossible to tell that we are not dealing with a pure wave. Traversing the electromagnetic spectrum from radio waves to light, and finally to X- and γ-rays, there is an increasing tendency for particle properties to dominate. In the infra-red portion of the spectrum the wave and particle aspect are of comparable importance, as discussed in Chapter One, while for γ-rays the particle properties are almost completely dominant.

Wave Mechanics and the Atom

The wave-particle duality of matter forms the basis of present-day atomic theory, and an outline of the way in which this awkward concept has been built into a solid and powerful theory is most important for any understanding of the atom. The reason that the wave aspect of particles is difficult to swallow, on first acquaintance, is that we have a strong feeling, based on experience, that we know all about particles. To state the obvious rather more precisely—a particle is a blob of matter which moves through space and of which we know the position and speed at all instants. This is just what Newton's laws of motion are all about; according to Newton, the motion of any body can be *exactly* calculated if we know the forces acting on it, and we can therefore predict the future track of the body through space. This is true to high accuracy for a tennis ball or a space vehicle, but for atomic particles the whole idea breaks down. We have to use *wave theory* to predict the direction in which an electron will bounce off a metal; in other words, the basic idea of being able to determine the future motion of a single particle through space is simply not true for atomic particles. After reflection some electrons are diffracted in certain directions and others in different directions. Calculation of these angles, using diffraction theory and the de Broglie wavelength, enables one merely to specify the *possible* tracks for one particular electron. Thus there is an absolutely basic uncertainty in nature; the future track of an individual electron cannot be calculated—we can only determine the probability that it will go in a certain direction. This concept, which differs radically from Newtonian ideas, is known as Heisenberg's Uncertainty Principle.

In the light of this principle we must seriously modify our view as to the nature of an actual particle. After all, if we cannot predict its path what right have we to picture it as flying through space like a tennis ball? It is just at this point that the real fallacy of the old corpuscular theory of light becomes apparent. Photons do *not* travel through space like a cloud of shot. It is impossible to say that a particular photon will be at a specified point at a predictable time; what can be stated is that each of a group of similar photons has an equal probability of arriving (and impacting as if it were a tiny particle) close to one of a series of possible points within a span of time also governed by a probability distribution.

Wave mechanics is the name given to the new method for calculating the behaviour of atomic particles, and the theory has a great deal to say about atomic structure. Before the advent of 'matter waves' an atom was believed to consist of a positively charged nucleus around which electrons circulated, rather like planets around the sun. A picture of hydrogen, the simplest atom, which consists, on the old view, of one positively charged proton as nucleus and a single electron, is shown in Figure 16. One of the troubles with this model is that we expect it to radiate electromagnetic waves continuously. A rotating charge generates a changing electric field somewhat analogous to the apparatus of Hertz (Figure 7 (a) Chapter 1), and one might expect it to launch waves in the same way. If this happened, the electron would lose energy and spiral in towards the nucleus, just like an artificial satellite losing height due to atmospheric drag in its orbit round the earth. However, we know that an electron can circulate about the nucleus indefinitely without emitting any radiation.

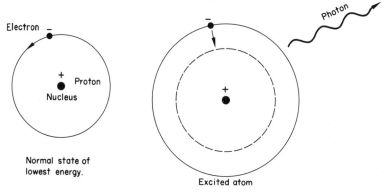

FIGURE 16 Orbiting electron model of an atom of hydrogen.

The radiation which atoms actually emit is easily studied by means of a diffraction grating, or a prism, which will separate out the different wavelengths. What we find is that the light, unlike that from a red-hot poker, consists of certain wavelengths only. The spectrum, or wavelength distribution, of light from sodium atoms (as in street lights) is compared with

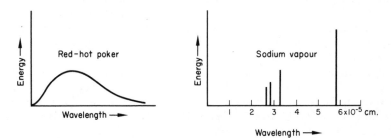

FIGURE 17 Energy spectrum of radiation from a hot body compared with that from sodium vapour.

that from a hot poker in Figure 17. While that from the poker covers—like sunlight—a wide range of wavelength, the sodium light all comes at one precisely determined wavelength. For sodium atoms disturbed rather more violently than those in a street light, many more wavelengths would be radiated, a few of which are also shown in Figure 17. These wavelengths are emitted only by sodium; different atoms have their own characteristic series of wavelengths or 'spectra', and this fact is of immense value in optical astronomy, since it affords a means of studying the composition of stars and galaxies (see Chapter 3).

On the orbiting model of the atom, it was supposed that the electron could gyrate at different distances from the nucleus and that a photon was emitted each time it jumped from one orbit to another with a smaller radius, as sketched in Figure 16.

As we have seen, however, the new wave-particle concept makes it rather meaningless to think of electrons actually moving in a circular path. We must calculate the probability of finding an electron somewhere using wave ideas. When this is done we obtain a much more realistic—though less precise—model of the atom, and the reason for the particular atomic wavelengths radiated becomes much clearer. It is no longer necessary to assume that only particular orbits are allowed by nature. The calculation of the distribution of matter waves near a proton (the hydrogen nucleus) is not difficult, but the answer is easier to visualise if we mention some simpler cases first. Take the string of a guitar, for example. The string is held down firmly at each end and, when it is plucked, it gives out a note which corresponds to the frequency of vibration; the greater the tension, the higher the pitch of the note. The actual motion of the string is small and difficult to see, but it is made up of simple oscillations, as depicted in Figure 18 (a). The largest movement is that in which the string goes up and down in unison along its whole length—this gives the fundamental note. But there are other oscillations, in which different parts of the string move in opposite directions. These correspond to notes higher than the fundamental and are called the 'harmonics'. It is the presence of a particular set of harmonics which makes the 'twangy' quality of a guitar so different from that of a bowed string in a violin. Now these oscillations can be thought of as waves on the string which are trapped between the two rigidly clamped ends, and the really important feature is that the wavelength of the trapped waves is determined by the length of the string. The waves must fit so that the string is motionless at each end.

A guitar string is, of course, one-dimensional, but the stretched skin of a drum can be thought of in much the same way. The up-and-down motion now has to be drawn as a contour map, and the fundamental and some of the harmonics are sketched in Figure 18 (b). The whole situation is much more complicated because the trapped waves can run around the centre, and from the centre to the edge. When a drum is struck there are so many harmonics present that we can scarcely detect the basic note of the fundamental oscillation. A bell is somewhat similar, although three-dimensional, but here the harmonics are controlled by skilful design so that a pleasing note is obtained.

Returning to the atom, the question of calculating what kind of particle waves can exist near the nucleus is exactly analogous to working out the vibrations of a bell, except that now we have to consider what wavelengths can be set up, or fitted into, a spherical volume of space surrounding the nucleus. However, by similar methods we can calculate the fundamental

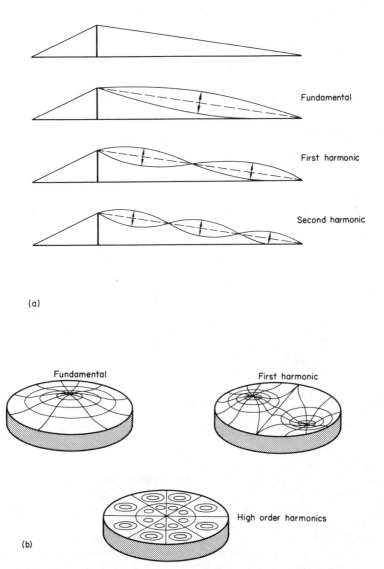

Fundamental

First harmonic

Second harmonic

(a)

Fundamental

First harmonic

High order harmonics

(b)

FIGURE 18 Possible modes of vibration of (a) a guitar string and (b) a drum.

oscillation and its many harmonics. Remembering de Broglie's relation, which states that the wavelength associated with a particle is shorter the higher the speed of the particle, it follows that the vibration of longest wavelength, the fundamental, specifies the lowest speed for the electron; the harmonics will then correspond to greater speeds. Actually it is better to consider the energy of the electron, rather than its speed, so that the vibrations represent a whole series of energies which the electron may have, the lowest being that appropriate to the fundamental. These are called the energy levels of the electron and refer to vibrations in a radial direction from the nucleus; some of these energy levels are shown in Figure 19.

The analogy of a bell cannot be pressed too far, one big difference between an atom and a bell being that a single electron can have only one

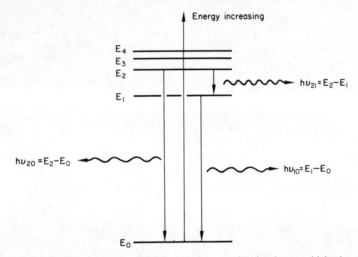

FIGURE 19 A few energy levels of the hydrogen atom showing jumps which give rise to radiated photons.

value for its energy at any instant; the harmonics cannot all be present together. More complicated atoms containing many electrons come closer to the bell model. The instantaneous energy of the single electron in the hydrogen atom depends upon the history of the atom. The energy can never be smaller than the lowest value shown in Figure 19, but collisions with other atoms or the gaining of energy from a photon can raise the electron to one of the higher levels. Once this has happened, the electron can lose its energy again by dropping from one level to another, and the energy lost is emitted as a photon, as shown in Figure 19. The frequency of the electromagnetic wave corresponding to a photon emitted when the electron jumps from a level E to a new level E' is given by $h\nu = E - E'$, where $h = $ Planck's constant. Thus the bigger the energy jump, the higher the frequency.

The heavier an atom is, the more electrons it contains and the more complicated becomes the calculation of its possible vibrations. It turns out that the most likely regions in which to find electrons take the form of shells around the nucleus, as illustrated in Figure 20. Only electrons in the outermost shell are able to change their energy easily. It is these electrons which are responsible for the emission of photons corresponding to light radiation. But when an atom is bombarded by electrons travelling at high speed an electron in one of the inner shells may gain sufficient energy to jump into a shell at a greater distance from the nucleus, or possibly to leave the parent atom altogether. When this happens, a vacant space is left in the inner shell, and this space will rapidly be filled by another electron 'falling' from an outer shell. A large energy jump is involved, and this corresponds to the emission of an X-ray photon. There are usually several ways in

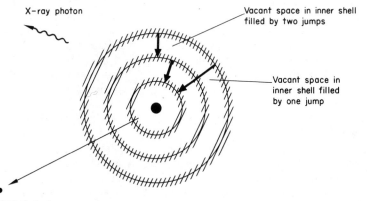

FIGURE 20 Shell structure of an atom derived from wave mechanics. X-rays are emitted when an electron 'falls' into a vacant space left in a shell closer to the nucleus.

which an outer electron may reach the inner shell. The greatest energy change occurs when an electron falls from the outermost shell, but smaller changes occur if the vacant space is filled by a succession of jumps between adjacent shells, as shown in Figure 20. X-rays of various characteristic wavelengths are therefore emitted from a given atom in much the same way as it emits particular light wavelengths.

The wavelengths of X-rays are considerably shorter than those of light waves, since the energy changes involved are so much greater, but even shorter wavelengths are generated when photons are emitted from the nucleus of an atom. The nucleus is a closely packed bunch of protons and neutrons which exists at the heart of an atom and which contains nearly all the mass of the atom. So far the structure of the nucleus is not fully understood, but there is no doubt that it possesses energy levels as does the electronic structure outside. When the nucleus is excited by collision with another nuclear particle, possibly derived from a high-energy accelerator, photons of extremely high energy and short wavelength may be generated.

These are known as X-rays, and their energy corresponds to the energy difference between shells in the nucleus. Again, particular nuclei are characterised by their own wavelengths, although at this end of the spectrum the photons have such enormous energies that the particle properties of the radiation are dominant.

From the previous discussion it is apparent that complex atoms containing many electrons are analogous to a bell, in that they may emit a whole range of characteristic wavelengths, or harmonics, when suitably excited. One small complication which has not been mentioned so far is an effect arising from the magnetic properties of electrons. To a good approximation we may regard an electron as a particle which simply carries a negative electric charge. More accurately, however, it also behaves as though it were a very feeble magnet. The magnetism arises because the electron spins like a top, and this property is known as electron *spin*. Because the electron acts like a magnet, the energy states of an atom will be changed slightly if the atom is subjected to a magnetic field. The effect is small, but it is important, as will be mentioned again in Chapter Six. Another effect of spin is to cause the electrons in each shell to 'pair off' in couples with spins aligned in opposite directions. When an electron has no partner it is called an 'unpaired' electron.

A further complexity to the possible oscillations of an atom arises when the atom is one component of a molecule. In a molecule several atoms are linked together and the whole structure may oscillate and rotate, in addition to effects within the atoms themselves. The far greater mass of a complete atom, as compared with that of a single electron, means that the photons radiated correspond to wavelengths longer than those of light waves. Molecular radiation usually falls in the microwave or infra-red region of the spectrum. The reason for this may be understood from de Broglie's relation (see p. 26). The particle wavelengths associated with atoms and electrons in ordinary matter are much the same, and de Broglie's relation then implies that the constituent atoms of a molecule must be moving far more slowly than the electrons of an atom. They therefore possess smaller energy and radiate less energetic photons.

The fact that atoms emit radiation which depends so much upon their individual structure means that a careful study of the radiation from, say, a distant star can tell us a great deal about the chemical composition and physical condition of the emitting region. Spectroscopic investigation thus forms one of the most powerful techniques in modern astronomy, as well as being important in many other fields.

So far attention has been given only to the emission of radiation, but atomic oscillations of all kinds are equally important with regard to *absorption*. It is an interesting fact that any body which strongly emits radiation of some preferred wavelength will also absorb strongly at the same wavelength. The reason for this is clear if we imagine an atom absorbing a photon and thereby causing an electron to jump to a higher energy level. Only if the incoming photon has just the right energy to raise the electron *exactly* to the higher level will the jump be likely to take place. When this is so, the process is called *resonance absorption*. A simple example of this occurs when white light is passed through sodium vapour; it is found that the emerging light has a spectrum in which the yellow sodium wavelength is largely missing. The composition of the atmosphere surrounding hot stars may be studied in this way, as mentioned in Chapter Three. The colour of different materials is caused in much the same way. Red paper looks red because the molecules of the dye have strongly absorbed the blue photons in white light.

It is perhaps surprising, after all this discussion of atomic vibrations, that it is possible to make white light at all. Why does white-hot metal radiate photons of all wavelengths, rather than photons characteristic of the metallic atoms concerned? The answer is not simple, and is related to the fact that when atoms are packed close together, as in solids, the electrons cannot be regarded as entirely belonging to one nucleus. In a sense, the collection of countless atoms must be thought of as a single super-atom. Calculations then show that the fundamental vibration occurs at such a low frequency that its harmonics are very numerous and closely spaced. This means, in effect, that the energy levels for the electrons in a lump of solid material are separated by such small gaps that virtually any level is permitted. Correspondingly, an electron can lose energy by any amount, and hence radiate a photon of any frequency. 'White' X-rays are emitted from a metal in the same way if the metal is bombarded with high-speed electrons.

Before leaving these awkward, but fruitful, concepts of particles and waves it is now possible to go back to the object sliding on a smooth table—the point where quantum ideas were first introduced. In order to calculate the least possible speed of the object, it was necessary to say how wide the table was. Admittedly this example was an extreme case, but the principles of wave mechanics must still be applied, and the fundamental vibration depends upon the size of the region in which we confine the waves. Hence the need for some definite dimension. If the object were allowed to fall off the table and move freely over the surface of the earth the minimum speed would be smaller still!

Light Waves and the Aether

Waves which we can actually see, rather than the intangible matter waves which need to be invoked to understand atomic behaviour, always travel *in* something. Sound waves are supported by air, and the compression waves can be made visible by suitable techniques; wave crests on water are obviously caused by motion of the liquid surface. In the previous chapter we saw how scientists imagined the 'aether' to be the medium which supported electromagnetic waves, although this concept raised a number of difficulties.

Because light waves travel so fast, it is not easy to design an experiment which could show up any motion of the aether, but in 1887 Michelson and Morley made an extremely accurate series of measurement. By an intricate arrangement of mirrors supported on a massive stone block floating in liquid mercury, they looked for a difference in the velocity of light when it travelled in two directions at right angles. Unless the aether is stationary relative to the earth, it must 'blow' past any fixed point on the earth, and the velocity of light will then be different along, and perpendicular to, the direction of the aether wind. In fact, no such difference was found, although the accuracy of the experiment was quite sufficient to have shown a discrepancy of an order of magnitude which would be caused by the velocity of the earth in its orbit. Other motions, such as galactic rotation or the velocity of the whole galaxy through space, would give rise to an even greater aether wind and a larger difference in the apparent velocity of light in two directions at right angles.

This result, one of the most important findings in the history of physical investigation, means one of two things. Either the aether is at rest, relative to the earth, or light waves are not supported by any medium at all. Now one cannot seriously suppose that the aether, which pervades the whole universe, is always moving at exactly the velocity of the earth, and the only way out is to suppose that the aether has no physical reality. The implications of this deduction were astonishing and led Einstein to develop his theory of relativity. As wave mechanics radically changed our view about matter, so the theory of relativity overthrew our preconceived ideas of space and time.

We are born and bred in a world where space and time seem absolute. Maps delineate the boundaries of countries and continents, while clocks display the continuous passage of time, and it seems ludicrous to suggest that miles and minutes are not quantities which are the same for all. Yet this is the essence of Einstein's theory. The Michelson and Morley experiment shows quite definitely that the velocity at which light travels through space is the same for all observers, regardless of how they may be moving themselves. A simple illustration brings out the apparent oddity of this situation. Imagine a train rushing through a station on a dark evening. The guard throws a switch and the lights come on in all the carriages simultaneously so far as the passengers are concerned. Now according to Einstein, a bystander on the platform will not see the carriages light up together. He will see the lights come on one after the other in sequence along the train. Of course, a certain time elapses before any light from the train reaches the spectator, but when this is allowed for the spectator would still think that the lights had come on sequentially. Naturally the effects are very small in this example but, for atomic particles which often travel at speeds near the velocity of light, relativistic phenomena are extremely important. So the aether can be dispensed with only if we are prepared to accept a world in which space and time are different for observers in motion relative to each other.

The pursuit of these ideas is a fascinating study which goes rather beyond the scope of this book. But it will be clear from the topics already discussed that investigations of the electromagnetic spectrum, and of light in particular, have proved to be a most fertile ground for extending our knowledge of the physical world.

THREE

LIGHT FROM THE STARS

Astronomy—the Basic Science

Astronomy has been called 'The Mother of the Sciences'—and rightly so, for the first truly scientific problems to be tackled by Man, at the dawn of history, concerned the motions of the heavenly bodies. To the eye, the stars appear to be attached to the inner side of a vast celestial sphere, which slowly and regularly rotates from East to West about an axis through the celestial poles—a rotation that we now recognise to be a reflection of the rotation of the Earth on its axis in the contrary direction. The 'fixed' stars seem to keep the same pattern on the celestial sphere year after year, but other bright objects, the 'wanderers' or planets, while sharing in the daily rotation, move against the background of the stars in apparently complicated gyrations. The problem that exercised scientists from the Greeks until the beginning of the seventeenth century, was this: how is it possible to represent the motions of the planets by some combination of simple motions, such as motion in a circle? This problem was finally solved, in the early years of the seventeenth century, by Johannes Kepler, who showed that the apparent motions of the planets were well represented by supposing that a planet revolved in an elliptical orbit with the Sun at one focus of the ellipse, the Earth itself being a planet moving in an ellipse about the Sun.

In another sense too, perhaps, astronomy has been a mother to the sciences. For science is more than a systematic description of the universe—it is also, and more fundamentally, an attempt to understand the universe in terms of general scientific laws. It was the desire to understand the reason for Kepler's Laws of planetary motion that led Newton to his famous 'Law of Gravitation', which states that the elliptical orbit of a planet arises from a balance between the tendency of the planet to move off in a straight line as a result of its speed and the gravitational force of attraction exerted on the planet by the Sun. To arrive at his solution, Newton had to develop methods of applying mathematics to physical problems, which methods still lie at the foundation of physical science.

Yet throughout all this long history of grappling with these problems of planetary motion, astronomers had given little, if any, thought to the nature of the light by which the planets and stars were visible to us. Why should they have concerned themselves with the nature of light? After all, they could simply look up into the sky on a fine night and see that the stars are there or, during the day, see the Sun. True, in order to formulate a problem of planetary motions, certain assumptions about light—for example, that it travels in straight lines—had to be made, but they were made implicitly, and not recognised as assumptions at all.

Newton himself, in the middle of the seventeenth century, was one of the pioneer investigators of the nature of light, on the basis of physical experiments in the laboratory. However, by Newton's day considerable practical experience in the making of spectacles and other simple optical devices had been acquired, and from this practical skill there emerged the telescope, first used for looking at the heavens by Galileo in 1610. At first the telescope was used simply as a magnifying glass, to see enlarged images of the Moon and planets. But, as a magnifying glass for use on the stars, the telescope must have disappointed its early users. It is true that the image of a star in a telescope does show a disc, but this disc is due to inherent imperfections of telescopes as optical devices and has nothing to do with the real disc of the star. Stars are so far away that their true discs are too small to be seen with even the largest telescopes—in a perfect telescope the image of a star would seem like a point.

Nevertheless, a telescope is useful for observing the stars, as it collects much more light than the unaided eye, and if all this light is focused at a point the point will be very bright. So a telescope increases the brightness of a star, and enables stars to be seen that are far too faint to be observed with the unaided eye. It is as a light gatherer, rather than as a magnifying glass, that the telescope has been of greatest use to astronomers.

Galileo noted that the Milky Way—a faint band of light that stretches in a ring right round the sky—could be partly resolved with his telescope into myriad faint stars. The stars in the sky were therefore not distributed through space at random like pepper from a pepper-pot, but formed some sort of system. Surprisingly enough, a century and a half was to elapse before anyone thought of using the telescope systematically to explore the distribution of stars in space.

Herschel's Attempt to Map the Galaxy

The first person to attempt such a survey of the stellar system, or Galaxy as we now call it, was William Herschel. To do this, Herschel had to devise a means of estimating the distance of a faint star, and he made use of the apparent brightness of the star in his telescope. He argued that if he assumed that all the stars in the sky were similar objects, then all the stars would give out the same amount of light. They would have, as we say, the same *intrinsic* luminosity, just as, for example, all 100-watt electric light bulbs give out the same amount of light. But the light of a star has to travel towards us and, in doing so, it gets spread out, and becomes fainter. For example, if there are two identical stars, one of which is twice as far away as the other, the farther star will appear only one-quarter as bright as the nearer one. So if he could measure the relative brightnesses of the stars Herschel could (he thought) obtain the *relative* distances of the stars, and so map out a scale model of the Galaxy. To find the actual size of the Galaxy would be more difficult, but if it were assumed that the Sun also was a typical star, and if the brightness of a star could be compared with that of the Sun (a very difficult task), then an estimate of the size of the Galaxy could be made.

Herschel had no sophisticated instruments to measure the relative brightnesses of the stars, and he could use only his eye as a light-recording instrument. His method was to focus on a particular star, and then alter the aperture of his telescope until the star just ceased to be visible to his eye. Since the amount of light entering a telescope is proportional to the area of its lens or mirror, Herschel was able to measure with some precision the relative brightnesses of the stars, although the Sun is so bright that he could not apply this method to it.

Of course, a modern astronomer does not use his eye for measuring the brightness of a star (except in very rare cases), as he has more precise and more sensitive light receivers. For example, if the light of a star, collected by a telescope, is allowed to fall on a photographic plate the blackening caused by the starlight is a measure of how much light fell on the plate. The blackening on a photographic plate can be measured accurately, at leisure,

FIGURE 21 The Galaxy as pictured by Herschel.

in the laboratory; furthermore, a permanent record of the star and its surroundings has been obtained. Alternatively, the starlight may fall on a sensitive photo-electric cell (as is used in a photographer's light-meter). The cell gives off an electric current proportional in size to the intensity of the starlight, so that the brightness of the star can be determined accurately by measuring the current given off by the photo-electric cell. A modern photo-electric photometer on a telescope is so sensitive that it can measure the brightness of a star too faint to be seen by eye using the same telescope.

Figure 21 shows the sort of picture that Herschel built up of the Galaxy. It represents a cross-section of the Milky Way. Seen from the top, the system would look more or less circular. So the Galaxy is in the form of a flat plate or disc, with the Sun close to the centre. When we look out along the plane of the disc we see lots of stars—the Milky Way—but when we look out across the plane of the disc we see very few stars. Nevertheless, there is something curious in Herschel's picture: why should there be so many indentations in the stellar system, and why should they all point towards the Sun? The fact is that Herschel's method failed to give him a true picture of the Galaxy.

Herschel's first assumption, that all the stars are similar and have the same intrinsic luminosity, is clearly a very doubtful assumption. Indeed, we now know that stars differ enormously, one from another, in their output of light. However, working on this assumption, Herschel was just as likely to have overestimated the distance of a star as to have underestimated it, so that, although his estimate of the distance of an individual star might have been quite wrong, this would not have caused any systematic distortion of his picture of the Galaxy, provided only that the distribution of stars of different intrinsic luminosity was much the same at different places in the Galaxy.

No, it was Herschel's second assumption—that the light of a star could be dimmed only by distance—that failed him. For, unknown to him, lying between the stars are clouds of interstellar dust (and gas). By terrestrial standards, interstellar dust clouds are extremely tenuous—if the Earth were actually inside a cloud of interstellar dust we would be unaware of the fact (without very refined astronomical measurements). But interstellar dust clouds are very large, and starlight has to pass through such clouds to reach us; in doing so much of the starlight may be absorbed. We can now see why Herschel found the Sun at the centre of his system—it is as though he were looking out through a fog which determined the distance to which he could see and, of course, resulted in his finding himself in the centre of the region visible to him. Where the interstellar fog is particularly dense, the stars at a given distance appear fainter, and so Herschel would have overestimated their distance. This explains the spikiness of his picture of the Galaxy.

Sadly enough, Herschel himself came to realise, without fully understanding the reasons, that his great scheme of mapping the system of the stars was a failure, and that the Galaxy was unfathomable even with his largest telescopes. The modern astronomer has succeeded in discovering the limits of the Galaxy, and far beyond, but to do so he has had to overcome the limitations of Herschel's assumptions. This has become possible through a thorough analysis of the nature of the light that the stars send to us.

The Information in Starlight

Light is a wave-motion in which the displacement takes place in a direction at right angles to the line of travel of the wave (see Chapter 1, p. 15). Suppose that Figure 22 represents a light wave. What questions can we ask about this light wave? First of all, we can ask what is its *intensity*—that is to say, how much energy in the form of light is crossing each square millimetre every second. The intensity of the light wave is related to the extent of the displacements from the 'zero line'. Secondly, we can ask in what plane are the light vibrations taking place—for example, in Figure 22 the angle θ would be a measure of the plane of vibration, or plane of *polarisation* as it is called. Thirdly, we can ask what is the *shape* of the light disturbance.

The shape of a light-disturbance may be very complicated, and so we need some means of specifying in detail what the shape may be. Consider the rather simple shape at the top of Figure 23. We could represent this shape as the sum of three even simpler, regular waves (called sine waves). Each of these sine-waves has a characteristic *wavelength*, λ, and an *amplitude*, a, measuring the extent of the displacement (see Chapter 1,

FIGURE 22 Example of a light wave from a star.

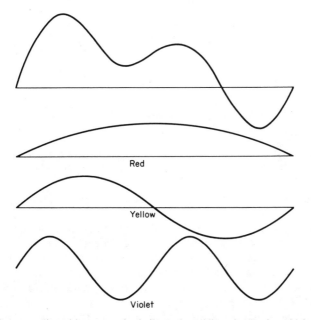

Red

Yellow

Violet

FIGURE 23 A complicated wave can be built up by adding simple sinusoidal waves of different wavelength. The different wavelengths constitute the spectrum of the disturbance.

Figure 9). We could thus specify the shape of the original wave by saying that it could be resolved into the sum of a wave of wavelength λ_1 with an amplitude of a_1 plus a wave of wavelength λ_2 with an amplitude of a_2 plus a wave of wavelength λ_3 with an amplitude of a_3. However complicated the original waveform was, we could always resolve it into simple sine waves in this way, although the more complex the original waveform, the greater the number of different wavelengths required to represent it.

We cannot observe the shape of a light-wave directly—but we *can* find out experimentally into which wavelengths it can be resolved, and with what amplitudes. The instrument for doing this is called a *spectrograph*, and it can be said to have been invented in the course of Newton's famous prism experiment (see Figure 24 (a)). Newton cut a hole in the shutter of his room and allowed sunlight to fall upon a prism. He noted that the prism produced a coloured band, or spectrum, from the white (or yellow) sunlight. The prism is, in fact, resolving the light wave into sine-wave components, and spreading the components out in a line—each particular wavelength λ corresponds to a particular *colour*, and the prism bends light waves of different wavelengths (or colours) through different angles.

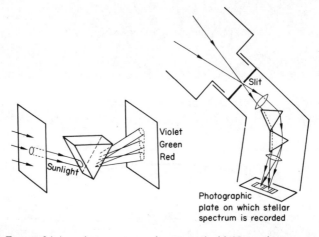

FIGURE 24 A modern spectrograph compared with Newton's apparatus.

If the light wave of Figure 23 were put through Newton's shutter, then his spectrum would have consisted of three circular images of the hole, one red, one yellow and one violet. A more complex waveform would give so many different colours that they would overlap in Newton's experiment. White light is the most complex waveform, containing all the visible colours. To overcome this problem of the overlapping of images, the source is made, not a circular hole, but a slit, the images of the slit in different colours lying side-by-side. (Figure 24 (b) shows the construction of a typical prism spectrograph at the end of a telescope.) So any point in the spectrum then corresponds to a specific colour, and the light intensity at that point is related to the amplitude of that particular wavelength in the original waveform. The spectrograph has analysed the light for us.

In practice, the spectrum is recorded on a photographic plate, the blackening at different places on the plate being measured to give the intensity of light of various colours. Now some astronomical objects do have spectra consisting of a few colours only. Such an object is the Ring Nebula in Lyra, and its spectrum could be photographed without using a slit source, for it would consist of a series of images of the nebula in different colours (i.e. different positions). If a slit is used, then the spectrum would consist of a few lines—images of the slit in different colours. For this reason such a spectrum is called a *line-emission spectrum*. I like to think of a spectrum as rather like a piano keyboard—with the bass notes corresponding to red light and the treble notes to violet light; but it is a keyboard with an infinite number of keys! A spectrum such as that of the Ring Nebula would correspond to the playing of a single chord, while the spectrum of white light would correspond to playing all the keys at the same time.

Objects which show a line-emission spectrum, like the Ring Nebula, are tenuous clouds of gas. To find out the nature of celestial objects, we must compare their spectra with spectra emitted by objects in the laboratory whose nature we know. Thus we would expect that if we can in some way put energy into a gas in the laboratory it should get rid of this energy by radiating a line-emission spectrum. Broadly speaking, this is so, the energy usually being provided either by a flame or by an electric spark, when the gas is said to be *excited*. Furthermore, it turns out that the particular colours that are radiated depend upon the chemical elements or compounds present in the gas (see Chapter 2, p. 29). Each element, so to speak, has its own characteristic 'chord' on the spectrum keyboard. When we see the same 'chord' in the spectrum of a celestial object we know that that particular element must be present—we can, in other words, make a chemical analysis of the celestial object, even although we cannot touch it.

On the other hand, a hot solid body, such as the filament of an electric-light bulb, gives out a *continuous* spectrum, in which all colours are present, although with different intensities. The relative intensities of different colours do, in fact, give a good indication of the temperature of the hot body. Imagine that you are heating a poker in a fire, in a darkened room. At first, when you withdraw the poker after only gentle heating, it will not be visible, for it will be radiating, not visible light but infra-red radiation, similar to light but of longer wavelength. As the poker slowly heats up, more of the energy is radiated in visible wavelengths, and the poker glows a dull red (see Figure 17). However, the poker has to be heated up to about 3,000° C before the brightest part of its emitted spectrum falls in the visible red. By now a real poker will have melted—but no matter; let us imagine that we have a poker that can be heated indefinitely. As it grows hotter the maximum intensity in the spectrum moves towards shorter wavelengths. At 6,000° C the maximum intensity is somewhere in the yellow-green. One must go to 10,000° C before the maximum intensity lies in the extreme violet region of the spectrum, and if the heating were continued further the maximum would pass into the ultra-violet.

This suggests that we might be able to tell the surface temperature of a star by looking at its colour. Stars do, indeed, give out continuous spectra, and we can tell the temperatures of stars from their colours. In fact, the difference in colour between one star and another is readily visible to the unaided eye, if a careful comparison is made. Blue stars are hot stars, with surface temperatures of perhaps 50,000° C and maximum energy intensity well into the ultra-violet. The reddest stars have surface temperatures of only about 2,000° C, and recently, using detectors of infra-red radiation, stars as cool as 800° C have been observed. The Sun, with a temperature of 6,000° C, is a relatively cool star.

However, the spectrum of a star is not just a continuous spectrum, like that of a hot filament. It is crossed by a number of *dark* lines. It has been found that the dark lines of such an *absorption line spectrum* correspond to bright lines in the spectra of recognised elements and compounds in excited gases (see Chapter 2, p. 35). For example, the spectrum of sodium shows two strong lines close together in the yellow, giving the well-known yellow colour of sodium-vapour street lamps. The same two lines occur in absorption (i.e. as dark lines) in the spectrum of the Sun.

The reason why the dark lines appear in the spectrum of a star, super-imposed upon a continuous spectrum, is that the gas which is giving rise to the dark lines lies above, and is cooler than, the layers of denser gas which give the continuous spectrum. The atoms in the cooler layers are absorbing more energy from the radiation falling on them than they are emitting. This process can be demonstrated in the laboratory. Suppose that you look at the spectrum of the flame of a Bunsen burner, into which some common salt has been put. The spectrum will be almost entirely a line-emission spectrum, with the two sodium yellow lines prominent, the salt (sodium chloride) having been broken up into sodium and chlorine atoms. Now allow sunlight to pass *through* the flame and into the spectrograph, and com-pare the spectrum with and without salt in the flame. The sunlight has come from a source much hotter than the Bunsen flame, and so, when sodium is present in the flame, the sodium lines in the spectrum appear darker.

And so we can learn from these dark lines not only about the chemical composition of the outer layers of a star but also about the way the temperature of the gases of which the star's atmosphere is composed changes as we go deeper into the star. Indeed, the temperature of the star's atmosphere very largely determines the character of the spectrum of the star. Why should this be? Well, let us consider again the experiment with salt and the Bunsen flame. The sodium lines appeared because, at the temperature of the flame, the molecules of salt were rushing around fast enough for a collision between two molecules to split them into their constituent atoms. At a much lower temperature the molecules would have been moving at a much slower speed, and a collision would not have been violent enough to split the molecules. The spectrum lines of sodium would not then have appeared, although somewhere else in the spectrum there would have been lines produced by the sodium chloride molecules. Raising the temperature of the Bunsen flame completely alters the character of the spectrum.

We now have two methods of measuring the temperatures of stellar atmospheres: observing the colour of the star, and observing the dark lines that appear in its spectrum. We would not expect these two temperatures to be the same, for the dark lines are formed in a cooler layer than the continuous spectrum, but we would expect some general relationship between the *colour* and *excitation* temperatures, as they are called, when we look at a large number of stars. In particular, the colour temperature should never be less than the excitation temperature. When we look at the spectra of relatively nearby stars we find a general relationship between the two colours of the type expected. However, when we look at more distant stars there is a general tendency for the colours to correspond to a lower temperature (by comparison with nearby stars) than would be expected from the excitation temperature. Sometimes very large anomalies in the colour are observed, and always in the same sense—the colour temperature is too low, the star is too red.

Interstellar Dust

This takes us back to the failure, due to interstellar matter, of Herschel's assumption that the light of a star could be dimmed only by distance. For, while absorbing (or scattering) the light of a star in all wavelengths, interstellar dust scatters blue light more efficiently than it scatters red light. The net result is that light which has passed through many dust clouds is redder than it was at the start of its journey, for it has lost more of its blue light than of its red light. This effect can be seen on a foggy day if a row of street lights is observed—the more distant street lamps, as well as being unduly faint, are also redder than the nearby ones. A modern astronomer must therefore allow for absorption by interstellar matter. By comparing the colour of a star with its line spectrum it is possible to calculate through how much interstellar matter the starlight has passed, and hence to estimate by how much the light of the star has been dimmed by interstellar absorption. Once known, this spurious absorption can be corrected for, and the true distance of the star found.

Interstellar dust can also be detected from the *polarisation* of starlight which has passed through it. The light from a star emerges from the star unpolarised—that is to say, it consists of vibrations whose planes are randomly distributed. Such light can be polarised by passing it through a sheet of Polaroid, for example. Polaroid consists of small elongated crystals embedded in gelatine, and lined up with their long axes parallel. This regular alignment means that light whose vibrations lie in a certain plane will be transmitted by the Polaroid more readily than light vibrations in a plane at right angles to it (see p. 7).

The essential property of Polaroid is that it should consist of elongated crystals suitably aligned. Now the grains of interstellar dust are elongated. Sometimes the directions of the grains are higgledy-piggledy, and starlight passing through a cloud like this would not be polarised. But quite often we find the starlight is polarised, showing that some mechanism must exist which has caused the interstellar grains to line their directions up like soldiers in a rank. It has turned out that our Galaxy has extensive magnetic fields which, by terrestrial standards, are very weak, but which are capable of causing the alignment of interstellar grains. From the measurements of the polarisation of starlight we may make maps of the interstellar magnetic fields.

Let us now turn to Herschel's first assumption—that all the stars are exactly similar, with the same intrinsic luminosity. The modern astronomer is better placed than Herschel in three ways. First, of course, he has larger telescopes and, what is much more important, he has, in the photographic plate and photo-electric cell, light-detecting devices much more sensitive than the human eye, and so he can use the light-grasp of his telescope more efficiently than could Herschel. In the second place the modern astronomer has available accurate estimates of the distances of nearby stars, obtained by a purely geometrical method. As the Earth moves around the Sun in its annual orbit the observer is changing his point of observation, and so, just as from a moving train the nearby telegraph poles appear to move against the background of distant hills, so will nearby stars appear to move against the background of more distant stars. The nearby star will appear to perform a small ellipse in the sky, over the period of a year, this ellipse being the same size and shape as the orbit of the Earth about the Sun would appear as seen from the star; the smaller the ellipse, the farther away is the star. By measuring this annual motion the distance of the star can be calculated in terms of the size of the Earth's orbit about the Sun. This method is called the method of *annual parallax*. In principle it is simple, but in practice it is difficult. Even for the nearest star, the greatest displacement of the star from its 'correct' position due to annual parallax is less than one second of arc (i.e. less than the size of a pin's head seen from a distance of a third of a mile). Astronomers, using the refinements of modern astronomical photography, can measure parallaxes down to one-hundredth of a second of arc (i.e. a pin's head seen from a distance of thirty miles).

The unit of distance used by astronomers is based upon the method of annual parallax, but it is more convenient here to use the *light-year*. This is a unit of distance, *not* time, and is the distance that a beam of light would cover in a year, travelling at a speed of 3×10^8 metres per second. The nearest star is about 4 light-years away from us, and the farthest star for which a direct distance has been measured by annual parallax is about 300 light-years. Within this distance of 300 light-years there are a few thousand stars, and the distances of most of them have been measured. For these stars we know both the apparent brightness in the sky and the distance, and so can calculate (making due allowance for interstellar absorption) their intrinsic luminosities. But 300 light-years is very small compared with the size of our Galaxy, which is some 100,000 light-years across. We must make use of a refined version of Herschel's method of distance estimating to push our survey out beyond the reach of geometrical methods of distance measurement.

Sorting out the Stars

Consider, then, one of these nearby stars. Its colour, or its spectrum lines, will tell us its temperature. Now, as a body is heated, two things happen to the radiation that it emits. The colour of the radiation changes, as we have already seen, and in addition the total amount of radiation emitted by the body increases. In fact, the amount of radiation (in all wavelengths) emitted by one square metre of a body can be calculated once its temperature is known. We can therefore calculate, for one of our nearby stars, how much energy of radiation is emitted by each square metre of its surface. We can also calculate, as we saw above, its *intrinsic luminosity*, which is the amount of radiant energy emitted by the whole surface of the star. A simple division then gives us the number of square metres in the surface of the star—from which we can obtain the *radius* of the star, since a star is spherical.

The stars vary considerably in size. The Sun, with a radius about one hundred times that of the Earth, is of average size. The star Antares, in the constellation of Scorpio, is a very bright star and is also a very red star, as can be seen at a glance with the naked eye. Since Antares is red, it must be a cool star, and each square metre of its surface gives out but little radiant energy. Nevertheless, it is a very bright star, and this must mean, even allowing for the possibility that Antares is near by, that its surface contains many more square metres than the surface of the Sun. Antares is a *red giant* star which, if placed where the Sun is, would engulf not only the orbit of the Earth but that of Mars as well. In contrast, consider Sirius B. This is a very faint companion of the bright star Sirius A, and it is visible only in a fairly large telescope. But Sirius B is a blue star, and therefore a hot star. To be so hot, and yet have such a small total output of light, means that Sirius B must be a small star, a *white dwarf*. In fact, Sirius B is only the same size as the Earth.

At first sight it might not seem surprising that stars vary so much in size—why should matter not have condensed into masses of all sizes? But we may learn something else about a star—namely its *mass*. Sirius A and Sirius B form a *binary star*—one moves in an orbit about the other like a planet around the Sun, and it is the selfsame force of gravitation, acting between the two stars, that causes this orbiting to take place. From careful observations we may find the size and shape of the orbit of Sirius B about Sirius A, and in this way calculate the force between them. This force depends upon the masses of the stars, and so these masses can be determined. There are many such binary stars in the sky, and whenever a star is a member of a binary system we can calculate its mass. The surprising fact emerges that while stars differ among themselves by a factor of a thousand million in luminosity, and by a factor of one hundred thousand million in volume, they differ in mass only by a factor of a few hundred: the range from one-tenth to fifty times the mass of the Sun embraces nearly all the stars in the sky. A star such as Antares must be very tenuous indeed—a room full of Antares would weigh less than 30 g—whereas a matchbox full of Sirius B would weigh 1,000 kg.

Perhaps these figures make one wonder whether there is anything wrong with the calculation of the radii of the stars. It so happens that, in specially favourable cases, as one member of a binary star moves about the other it periodically gets between us and the other star. Such a system is called an *eclipsing binary star*, and from observations of such stellar eclipses a direct measurement of the radii of the stars is possible. The answers always agree well with those obtained by calculation from the colours and luminosities of the stars.

The atmospheres of Antares and Sirius B must differ enormously in gas pressure—that of Antares must be as tenuous as a good laboratory vacuum, while that of Sirius B must be pressed hard down by the strong gravitational pull of the small but massive star. These differences of pres-

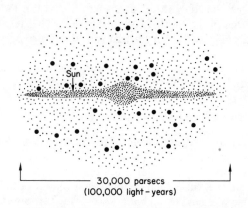

30,000 parsecs
(100,000 light – years)

FIGURE 25 (a) A recent model of our Galaxy.

sure have their effect on the spectra of the stars. Broadly speaking, while an increase of pressure tends to push atoms together to form molecules. An increase of pressure therefore mimics, to some extent, a decrease in temperature. However, the effect of pressure differs from one molecule to another, or from one atom to another, and small but observable differences between the spectra of giant and dwarf stars exist, certain spectral lines being very sensitive to pressure. It is therefore possible to recognise that a star is a dwarf or a giant simply by looking carefully at its spectrum, without knowing its distance (and hence its intrinsic luminosity). Unfortunately, it is not possible to calculate reliably from theory alone what the spectrum of a star of a given size, mass, and temperature would be like in detail, but one can use the stars which are relatively close to us and whose distances and intrinsic luminosities are known to find the characteristic properties of spectra corresponding to known luminosities. These luminosity criteria can then be applied to more distant stars, on the reasonable assumption that stars near to us are in no way peculiar, but are similar to more distant stars of the same types. Knowing the apparent brightness of a star, and estimating its intrinsic luminosity now from its spectrum, we can calculate the distance of the star—and we can do this for any star as long as it is bright enough for us to photograph its spectrum.

It is by using this method of *spectroscopic parallax*, as it is called, that the Galaxy has been mapped out. Figure 25 (a) shows a representation of what the Galaxy looks like from the edge, so to speak. Most of the stars, and especially the bright blue stars which are the most easily observed ones, tend to be concentrated towards the plane of the Galaxy, but the Galaxy has a halo, much less concentrated to the plane, in which the brightest stars are red, not blue. The large dots represent the globular star clusters—collections of thousands of stars, held together by their mutual gravitation—and these seem to be associated with the halo rather than the disc of the Galaxy. Seen from the top, so to speak, the Galaxy would look more or less circular, but with a pronounced spiral structure. Figure 25 (b) shows the Whirlpool nebula—this is another galaxy, containing, like our own, some hundred thousand million stars. We see it by geometrical chance from the top, and it gives us an idea of what our own Galaxy would look like from far off. Our own Galaxy even has two satellite galaxies, like the one with the Whirlpool nebula, which would look, to the naked eye, like pieces broken off from the Milky Way. They are called the Clouds of Magellan, but are visible only from the southern hemisphere. The spiral patterns common in galaxies being marked out by the bright blue stars, show up well on photographic plates, which are more sensitive to blue than to red light. As the bright red stars are usually in the halo, they do not help us to see spiral structure to the same extent.

Theoretical Considerations

The object of science is not merely to describe the universe, however; it is also, and more importantly, to understand it. Can we understand how this Galaxy of ours holds together? We have seen how the force of gravitation between the Sun and the planets holds the solar system together, and how this same force holds double stars and star clusters together. Let us apply the idea of gravitation to our Galaxy, supposing that each star attracts every other star. The centres of galaxies are bright and, as a first shot, we might suppose that nearly all the mass of the Galaxy is concentrated at its centre. The orbits of the stars (including the Sun) about the centre of the Galaxy will then be similar to the orbits of planets about the Sun. Just as Mercury moves faster than Venus, and Venus faster than the Earth, so will stars nearer to the centre of the Galaxy move faster than stars farther away. In Figure 26 the Sun is moving round the centre of the Galaxy, due to gravitation, in a circular orbit. The stars X_1 and X_2 are moving round more quickly, so that the distance between the Sun and X_1 is decreasing, while the distance between the Sun and X_2 is increasing. On the other hand, X_3 and X_4 are moving more slowly than the Sun, and so the

FIGURE 25 (b) The Whirlpool nebula.

distance between the Sun and X_3 is increasing, while that between the Sun and X_4 is decreasing. If we could measure the line-of-sight velocities of these stars we could confirm the differential rotation of the Galaxy, and furthermore, estimate the mass of the whole Galaxy.

The spectrograph comes to our aid again here. If a source emitting light is approaching the observer, then the light waves are, so to speak, squashed up, their wavelengths becoming shorter than normal, whereas if the source is receding from the observer the light waves will be pulled out and their wavelengths will be longer than normal. This is the Doppler effect, which was explained in Chapter One. In the case of light, shortened wavelengths correspond to light that is farther into the blue, and increased wavelengths to light that is farther into the red. If, then, we see a stellar spectrum in which all the expected absorption lines are there, in the right relative positions, but the whole spectrum is shifted slightly to the blue, we

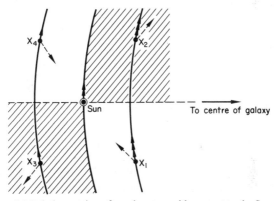

FIGURE 26 Relative motion of nearby stars with respect to the Sun.

may say with confidence that the star is moving towards us. Similarly, a slight red shift would be interpreted as a speed of recession in the line of sight. The amount of the shift would, in either case, give the speed of the star. (It should, perhaps, be emphasised that a star whose spectrum has a blue Doppler shift is not necessarily any *bluer*—it may even be redder, for while some radiation that, without the Doppler shift, would have been in the invisible infra-red is taken into the visible, other observable radiation of extreme violet colour is taken into the invisible ultra-violet. But in any case, for stars, the actual Doppler shifts are very small.)

Hence, in Figure 26, the spectra of stars X_1 and X_4 should show a violet shift, while those of X_2 and X_3 should show a red shift. Stars will not, of course, move in exactly circular orbits, but, taking a *statistical average*, the stars closest to us in the shaded quadrants should show a preference for red shifts, and those in the unshaded quadrants a preference for blue shifts. This is exactly what is found. We have extended the range of application of the law of gravitation, and found that the Galaxy has a mass equivalent to some hundred thousand million stars of the same mass as the Sun. Having been so successful with our theory of gravitation on the stars, we might try to observe the velocities of interstellar gas clouds, and apply the same type of analysis. It turns out that, in the case of gas, forces other than gravitation must be important, namely magnetic forces. The spiral arms of a galaxy are related to magnetic fields, and cannot be explained solely in terms of gravitational forces. But interstellar gas is best observed by radio astronomers and this will be discussed in Chapter Four.

While we have succeeded, in some degree at least, in understanding the structure of the Galaxy, we have said nothing about the structure of a star. Here a study of the light from a star can be of no help. It can tell us what the surface conditions of the star are like, how massive it is and how large it is, but it cannot give us any direct information about the inside of a star. Light is there in plenty, but it is absorbed within the star before it has a chance to escape. Indeed, what we call the 'surface' of a star is simply that region of its atmosphere from which the radiation has some chance of escaping unabsorbed into space, eventually to be collected by our telescopes. For our knowledge of the inside of a star we have to fall back on theory, and ask the question

'If I take a given mass of gas and allow it to condense under its own self-gravitation, will it at some stage start to generate energy and shine, and if so will it look like a real star?'

Using laws of the behaviour of matter found from terrestrial physics, a quite unambiguous answer is found: most of the stars are almost wholly composed of hydrogen, and they generate their energy through reactions between atomic nuclei at the high temperatures (tens of millions of degrees) at their centres. To be able to generate this energy, and have a surface temperature great enough for visible light to be emitted, the gas condensation must be at least as massive as the smallest stars, while if it is much more massive than the most massive star it would be too hot at the surface to remain stable. In fact, the hottest stars that we can see are continually losing their atmospheres into space. Theory thus explains why the stars come only in certain combinations of mass, luminosity and radius.

The theory of stellar structure can also trace the evolution of a star, from being like the Sun, through a red giant phase and finally to a white dwarf. It tells us that the bright blue stars of the galactic plane are relatively young stars, less than 10 million years old, having only 'recently' condensed out of interstellar matter. On the other hand, the red giants of the halo are perhaps as old as ten thousand million years. From the combination of the theory of stellar evolution and observation of stellar motions in the Galaxy, a picture of its evolution is being built up. It seems that it condensed out of *intergalactic* matter some ten thousand million years ago. During the rapid initial phases of the contraction the Galaxy was more or less spherical, and it was at this time that the stars of the halo were formed. During later stages of the collapse the material remaining became condensed into a disc, and out of this disc the later generations of stars were formed, a process that will continue until all the interstellar matter is used up.

We can see something of this process of galactic evolution by looking at other galaxies—for we have seen already that our own Galaxy is not the only one. Indeed, within the grasp of the 200-inch telescope are many hundreds of millions of galaxies. We find that some galaxies show a great profusion of spiral arms and bits of spiral arms—such a profusion in fact that the galaxy as a whole gives the impression of confusion. Such galaxies are called 'irregular' and, as they abound in bright blue stars and interstellar gas and dust, they must be young. Our own galaxy is middle-aged, showing some definite spiral arms, and with enough interstellar matter to keep star formation going at a steady rate in the disc. Elliptical galaxies are old galaxies. They look like galactic haloes without discs, for all the interstellar matter has been used up, and all the stars have passed into the red giant phase or beyond, and no bright blue stars are left to mark out the disc. Such will our own Galaxy become in due time.

As planets move round stars, and stars move round the centres of galaxies, do galaxies themselves perhaps move in some ordered fashion about the centres of supergalaxies? Astronomers have, of course, tried to extend into intergalactic space the techniques of distance measurement used within our own Galaxy, but no clear answer has emerged to the 'supergalaxy' question, although it is obvious that galaxies do show a tendency to form clusters.

The main difficulty in mapping out the system of galaxies is that we cannot observe the spectra of individual stars, even in nearby galaxies, largely owing to the overlapping of the spectra of so many stars. We can, however, observe individual star images, especially if the star varies its brightness. Variable stars of many types are common in our own Galaxy, and some at least show a close relationship between the period of the variation of their light and the total amount of light emitted. Hence, if we can recognise such stars in other galaxies from their light variations we can estimate their intrinsic luminosities and, from their apparent brightnesses, calculate their distances.

But, as before, we have to assume that variable stars in other galaxies are exactly similar to variable stars in our own. For nearby galaxies this may be reasonable, but we must remember that, when we look out into space we are also looking back in time. Even from the nearest galaxies, light has taken $1\frac{1}{2}$ million years to travel to us, and the light from distant galaxies was emitted thousands of millions of years ago, times at least as long as the time-scale of our galactic evolution. Is it reasonable to suppose that, throughout such long periods of time, no changes in the essential characteristics of stars have taken place? As yet, we do not know.

If the Universe itself were not evolving, perhaps it would be reasonable to suppose that the basic laws of physics would not have changed, but there is one difficulty in the way of this view. The spectrum of a distant galaxy is a sort of average spectrum of an average star in the Galaxy, with the lines somewhat blurred because of the motions of stars within the Galaxy. However, when we look at the spectra of galaxies we find the spectra always show a red shift, and the fainter the galaxy (i.e. the farther away the galaxy is), the greater is the red shift. The natural interpretation of these red shifts is as Doppler shifts. The whole system of galaxies seems to be expanding—but there is no centre of repulsion. Since the speed with which any two galaxies are separating from each other is strictly proportional to the distance between them, every galaxy sees the universe expanding symmetrically about it. At first the Doppler interpretation of the red shifts was accepted without question, but some doubts have crept in now that red shifts corresponding to a recession speed of four-fifths of the velocity of light have been measured. So far, no satisfactory alternative explanation has been put forward, however. If the Universe is really expanding, then *either* the average distance between galaxies is continually increasing (i.e. the Universe is in some way evolving) or new galaxies are continually formed to fill in the space left by the expansion, these new galaxies being formed out of matter that appears from nowhere. This is why the 'steady-state' theories are often called theories of 'continual creation'.

The idea of continual creation is not as much in conflict with modern physics as might be thought, for the creation and destruction of matter, and its transference to and from energy, are commonplaces of the physics laboratory. Indeed, if the steady-state theory were to prove true there would be a remarkable joining together of cosmology (the study of the very large) and atomic physics (the study of the very small), for the rate of creation of matter would be intimately connected with the structure of the atomic nucleus.

However, we must not forget the other possibility, namely that the Universe is expanding and evolving, difficult as this would make it for our assumption that distant galaxies were much like nearby ones. If this assumption fails, then we have no sure means of mapping out the system of galaxies. The choice between the steady-state theories and the evolutionary ones is therefore crucial. The decision may well come—indeed, perhaps it has come already—not from the optical astronomer but from the radio astronomer who, observing the radio waves from space, has detected objects farther away from those recognised by optical astronomers. Indeed, the radio astronomers have helped the optical astronomers to recognise a new class of very distant objects, the quasi-stellar galaxies, which had hitherto been confused with stars in our own Galaxy. But this is a story for a radio astronomer to tell, and will be taken up again in Chapter Four.

We have seen how a study of the light sent to us by the stars has enabled us to map our own stellar system and to understand its structure. We have discovered the chemical and physical nature of the atmospheres of stars and, with the help of theoretical physics, probed deep into the stars' interiors where nuclear energy is being generated. We have looked at our Galaxy in detail and, once more with the help of theory, found out how it has been evolving over thousands of millions of years. Finally, we have pushed our survey out, albeit with ever-increasing uncertainty, to the far-distant galaxies where, it would seem, velocities of recession from us are verging close to that ultimate barrier, the velocity of light. All this has been discovered from observations made in a narrow band of wavelengths, including the range to which the human eye is sensitive.

Just as the atoms of the cool outer layers of a star can absorb light, so can the atoms of the Earth's atmosphere. We can recognise dark lines common to all stellar spectra, which are due to atoms in the Earth's atmosphere. These are only a nuisance, because we know where they are. The real frustration is that we can use only the limited band of wavelengths that penetrate to us through the Earth's atmosphere. If we go but a short way into the ultra-violet, or somewhat farther into the infra-red, the Earth's atmosphere becomes completely opaque. The radiations that are absorbed carry important information about the stars and other stellar systems. It is exasperating to the astronomer to realise that radiation which may have taken a thousand million years to cross space to the Earth is lost completely in less than a thousandth of a second at the end of its journey through being absorbed by the Earth's atmosphere.

FOUR

THE RADIO WINDOW INTO SPACE

The Discovery of Radio Signals from Space

Conventional astronomy, making use of light waves, provides a wealth of information about our surroundings in the Universe, but we are still far from understanding a great many fundamental issues, such as the life history of galaxies or the creation of the Universe. Dr. Ovenden explained in the last chapter how frustrating it was that the information carried across space by light waves, possibly for some thousands of millions of years, was lost due to the action of the earth's atmosphere when almost within our grasp. Great strides are being made in the novel art of space-astronomy (observations made from artificial satellites beyond the atmo-

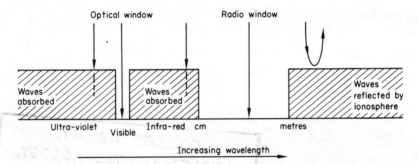

FIGURE 27 Optical and radio windows through the atmosphere.

sphere), but it will be a difficult and costly programme to construct observatories in space which contain telescopes comparable in size to those which already exist on the ground. It is not generally appreciated, however, that there is an extensive 'window' through the atmosphere at radio wavelengths. This window extends, as illustrated in Figure 27, from short wavelengths of a few centimetres to longer radio wavelengths of 20 metres or so. In the range between the optical window and the radio window the incoming waves react strongly with atoms and molecules in the atmosphere, while radio waves of too great a wavelength are reflected back into space again by the ionosphere. The latter is a belt of ions (atoms shorn of some of their electrons by ultra-violet photons contained in sunlight, as mentioned in Chapter 1) concentrated at a height of 100–300 kilometres. This belt is most pronounced in the daylight hemisphere, but the decay of the ionized layer during the night is not sufficiently rapid for the ionosphere to disappear completely before the morning.

Looking back on the history of radio astronomy, it is astonishing that serious use of the radio window was not made sooner. As long ago as 1932 it was realised that radio waves from outer space were penetrating to the earth's surface. This discovery was made by Karl Jansky, an engineer employed by the Bell Telephone Co., U.S.A., to investigate sources of radio interference. Jansky found that 'interference' of a special kind, known these days simply as radio *noise* and similar in character to the hiss of a radio set with the volume control turned fully up, was being emitted by the Milky Way. To be certain of this result required a fair amount of patience. The positions of the stars in the sky are governed by sidereal time, which gains just 3 minutes 56 seconds each day as compared with solar time. Jansky made recordings of the interference and, after a few weeks, was able to show that the time at which it became most pronounced occurred earlier each day in exact accord with sidereal time. This showed that the interference was not man-made, and it did not take Jansky long to realise that his aerial was directed towards the Milky Way when the interference was at its maximum.

Jansky's discovery caused a small stir in the Press of the day, but its potentialities for astronomical research went almost unnoticed. The only serious follow-up was made by a radio amateur, Grote Reber, who constructed a small radio telescope in his garden and carried out a survey of the radio emission over the whole sky. Possibly if the Second World War had not intervened, more thought would have been given to these pioneer studies.

The next stage occurred in 1942 and involved peculiar radio waves which, for a time, blotted out a British army radar network. J. S. Hey, engaged in operational research, analysed the reports as they came in and eventually concluded that the interfering signals had originated in the sun itself. He then noticed that a large sunspot was present on the disc and suggested that the radiation was, in some way, connected with it.

In 1945–46 many scientists who had been engaged in wartime research returned to their universities and serious attention was, for the first time, devoted to a study of these radio signals from outer space. Particularly notable were the groups led by Ryle at Cambridge, Lovell at Manchester and Pawsey at Sydney. Initially, work was concentrated on the Sun, and it was soon apparent that, with sensitive equipment, radio signals from this source could be received during all the daylight hours. Fantastic enhancements, sometimes as large as a millionfold, were also noted when particularly active sunspots crossed the disc, and this confirmed the wartime result. Solar radio astronomy has now developed into a major study at numerous observatories throughout the world, but the early work also pointed the way to even more exciting avenues of exploration.

As the first radio telescopes, which amounted to little more than modified radar receivers connected to simple aerial arrays, were gradually improved, it was found that space contained other radio emitters besides the Sun and the Milky Way. No visible stars could be found to correspond in position with these other sources, and they were initially called 'radio stars'. Some years later, following accurate measurements of position made by Graham Smith at Cambridge and by Bolton at Sydney, the 200-inch optical telescope at Mount Palomar in California was directed towards one of these mysterious radio stars. The photographic plates which were then made showed, for the first time, something of the power of the new vision which radio astronomy was to provide. One radio star, which is in the constellation of Cygnus and which is one of the most intense radio emitters, was unmistakably coincident with a galaxy of strange appearance lying near the limit of detectability of the 200-inch telescope. Apart from the odd character of this galaxy, it was found that it was situated at a distance exceeding 500 million light years, a distance which is beginning to become comparable to the scale of the universe itself.

Now the importance of making observations at very great distances is that they provide, in effect, a glimpse into the past. The radiation from the galaxy in Cygnus takes over 500 million years to cross space, so that radio waves reaching us *now* give us a picture of the galaxy as it *was* when they were emitted 500 million years ago. By observing even fainter radio galaxies at still farther distances it is therefore possible to look back in time and hence to learn something of the history of the entire universe— perhaps even to investigate its very origins. Before describing the extent to which these exciting possibilities have been realised, it is instructive to see how radio telescopes work and in what way they differ from conventional telescopes.

Radio Telescopes

Ordinary telescopes which make use of light waves are really nothing more than a logical extension of the human eye, and they have to fulfil two basic requirements. They must be sensitive to small amounts of light so that faint light emission from distant stars or galaxies can be detected; they must also be able to reveal fine detail such as the craters on the moon, or the exact form of galactic structure. Both these requirements depend, as might be expected, on the size of the telescope, and it is important to see just how this comes about. Since the light from a distant star travels in the form of waves, the radiation is spread out uniformly in space and the quantity of energy (or the number of photons per second) which falls on a certain point of the image is directly proportional to the area of the lens at the front of the telescope. Of course modern telescopes use a reflecting mirror rather than a lens, since this is easier to construct and less light is wasted by absorption in the glass; the general principle is the same as illustrated in Figure 28. For the human eye, the light-collecting area is only about one-hundredth of a square inch, but the giant telescope at Mt. Palomar has a collecting area of about 40,000 square inches. Since the energy in a light wave varies inversely as the square of the distance from the source, we deduce that with this telescope it is possible to see about two thousand times deeper into space than with the unaided human eye.

But light collecting is only part of the story, The detail which can, in principle, be seen using a telescope is governed by the bending or diffraction of light waves at the edge of the lens or mirror (see p. 19). This diffraction causes even 'point' sources, such as distant stars, to be blurred out into a patch of definite size. Thus, however large a magnification is used in the telescope eyepiece, or however great an enlargement is used when printing the photograph, the fine detail in the picture is ultimately limited by diffraction. This limit is called the *angular resolving power* of the instrument, and again it depends upon the size of the telescope. In astronomy it is convenient to specify resolving power in terms of the smallest angle which can be seen, e.g. the angle subtended by two stars when they are so close that it is only just possible to discern that there are

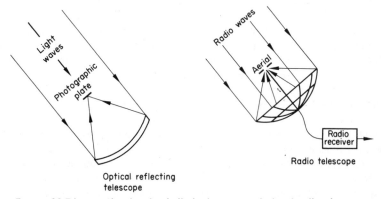

Optical reflecting
telescope

Radio telescope

FIGURE 28 Diagram showing the similarity between optical and radio telescopes.

two stars and not one. This angle is given by the ratio wavelength/diameter of the objective lens or mirror. Referring back to the 200-inch telescope again, this ratio is about one in 10 million, which corresponds to the angle subtended by the opposite sides of a penny at a distance of 150 miles.

The performance of a telescope is thus specified by its collecting area and its resolving power, and we might expect the same general considerations to apply to radio telescopes. Now radio waves are typically about 1 million times longer than light waves, and it is immediately obvious that to construct a radio telescope with the same angular resolving power as a good optical telescope is quite impossible—a radio telescope comparable to the 200-inch optical telescope would measure over 5,600 miles across! Before describing some ingenious attempts now being made to overcome the serious limitation of inadequate resolving power another difficulty—the impossibility of using photography with radio waves—must be discussed.

When a photographic plate is exposed at the focus of a telescope a permanent record is made of the light which entered the telescope from different parts of the sky. In a similar way the brain takes note of images formed on the retina of the human eye. Unfortunately the only way of detecting radio waves is to erect some kind of aerial and then to amplify the received signals electronically. The great disadvantage of this method as compared to photography is that it is impossible to spread many detectors across the focal plane of a radio telescope in order to obtain an instantaneous 'picture' of the radio sky. The only way to gain a radio picture is for the radio telescope to scan the heavens so that a record is made, sequentially, of the strength of the radio signals coming from each different part. A similar kind of procedure is employed in television, where the received picture is actually made by varying the intensity of a bright spot which moves over all points of the screen.

Returning now to the problem of resolving power, it is quite obvious that radio telescopes must employ very large reflectors, as diffraction otherwise leads to the production of exceedingly blurred pictures. It might be thought that the difficulties would be minimised by operating at a short wavelength, and this is partly true, but the radio emitters in space generally radiate more strongly at the longer wavelengths, and this means that attempts to use too short a wavelength will suffer from inadequate signal strength. In any case it is frequently desirable to make observations over a wide range of wavelength, since a study of the different radio 'colours' can yield most valuable clues concerning the physical processes taking place in distant galaxies.

Certain types of radio telescope, such as the giant 250-foot reflector at Jodrell Bank, are exact counterparts of optical reflecting telescopes. The only difference is that the surface of the reflector does not appear shiny. Of course, it is 'shiny' at the radio wavelengths for which it was designed and, in order for this to be so, the deformations of the surface must be small in comparison with the wavelength. The Jodrell Bank reflector is made of steel plates welded together, and the surface is constructed to be accurate to two or three centimetres. The shortest wavelength which can be used is therefore about 30 centimetres. Reflectors are frequently constructed using an open mesh which has the advantage of a smaller wind resistance and better drainage in wet weather. It is worth pointing out here that radio telescopes are unaffected by most weather conditions. Clouds are quite transparent except at the shortest wavelengths (less than about 10 centimetres), and so radio astronomers do not worry a great deal about the weather and can generally make useful observations for twenty-four hours each day.

A schematic diagram of a conventional radio telescope is shown in Figure 28. Until the advent of the computer era some ten years ago the records (corresponding to optical photographs) of the strength of the received signals were made on paper charts using pen recorders. Nowadays it is usually more convenient to analyse the signals in a computer, and so records are also taken, using some standard binary code, on a punched paper tape.

Anyone who has been near a large radio telescope will appreciate the complexity and magnitude of the structure needed to support the reflector so that it does not bend from its required shape. The Jodrell Bank instrument is still the largest of its kind, although a more accurate reflector of comparable size exists at Parkes in Australia. A project for a 600-foot reflector was abandoned in the U.S.A. at quite a late stage—the concrete foundations had already been poured before it was realised how costly it would be to incorporate automatic servo-mechanisms to keep the instrument aligned. Apart from the cost, however, it is clear that radio telescopes of conventional design cannot be made very much larger than a few hundred feet in diameter, and this fact led to the introduction of some radically new designs.

Foremost in the field of unconventional instruments is the system known as 'aperture synthesis' which has been pioneered by Ryle at Cambridge. This technique has proved so successful that a radio telescope comparable in power to a reflector one mile in diameter has been in operation at Cambridge for some years. The principle of 'aperture synthesis' is really very simple, but its translation into a practical system demands sophisticated electronics combined with computing facilities which can be provided only by a modern, high-speed computer.

To understand the basis of the synthesis process it is convenient to imagine some giant reflector—so large that it could not possibly be built. The waves falling into the bowl would be concentrated to a single point at the focus where the collecting aerial is situated. Now suppose that only a small portion of the reflector is actually in place, this element being movable so that it can occupy any position in the imaginary giant reflector. Then, by storing the radio signals, it is possible to measure, step by step, the waves which would have fallen upon the focal point of the equivalent giant reflector. In this way it is possible to simulate exactly the operation of radio telescopes far larger than can be built by present-day engineering techniques.

In practice, two small elements of the equivalent reflector are required, rather than one, because it is vital to measure the phase of the incoming waves, and this is most simply done by observing the phase difference between the two elements. One element is then fixed, to establish a constant phase reference, and the other is moved, relative to it, in order to carry out the synthesis. The radio signals at the output of the common receiver are recorded on a punched paper tape and, when all positions of the movable element have been covered, this data is fed into a computer. Calculations are then carried out which correspond to adding the reflected waves at the focus, and the result is a picture of the sky the same as would have been obtained by the equivalent giant radio telescope.

Various methods have been used in practical applications of synthesis. The 'one-mile' instrument at Cambridge employs three 60-foot reflectors, two of which are on foundations in the ground while one is mounted on wheels so that it can move along a 2,500-foot length of railway track. The rails are aligned east–west and at each of various positions along them recordings are made for twelve hours. During this time all the reflectors are driven to follow the same fixed point in the sky, just as in long-exposure observations of a star or galaxy using an optical telescope. As the earth rotates, the three reflectors, viewed from the sky, revolve about each other. Taking the rail-mounted reflector as a reference, the other two move round it and sweep out annular rings which are equivalent to the corre-

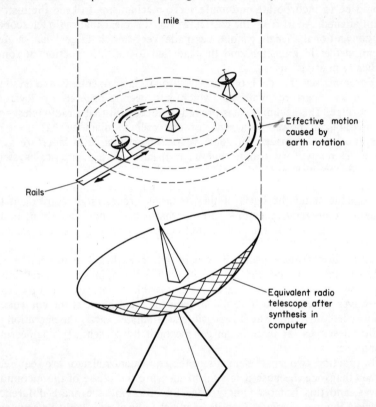

FIGURE 29 The principle of aperture synthesis as employed by the one-mile radio telescope at Cambridge.

sponding sections of a giant instrument. When all positions on the rails have been occupied the synthesis is complete. Only two reflectors are really needed, but the use of three enables the synthesis to be carried out in half the time, in addition to halving the required length of railway track. Since the observations for a full synthesis take two months to complete, a factor of two in observing time is a great advantage. The Cambridge radio telescope is shown schematically in Figure 29, while the actual structure and a radio map obtained with it are illustrated in Figure 30.

Because radio telescopes using synthesis can be extremely powerful while only requiring reasonably modest components, it is obvious that they are immensely important for making the deepest explorations of extragalactic space. However, it should be mentioned that they are not suitable for certain tasks, since they require that the astronomical objects under investigation do not change during the lengthy observing period. They would not, for example, be useful for studying stellar explosions known as supernovae, where significant changes can occur in a matter of days or weeks.

The Radio Sky

If we could actually 'look' through a radio telescope, perhaps the most surprising feature of the radio sky would be the brightness of the Milky Way. Also, since the atmosphere is completely transparent to radio waves, there would be nothing corresponding to the blue sky to which we are accustomed during the day. Indeed, there would be little difference between night and day. Superimposed upon the broad band of the Milky Way, we should notice the radio galaxies, by far the most common objects in view, together with the radiation from past explosions of stars within our own galaxy and glowing clouds of hot gas near particularly bright stars. If it were daytime the radio sun would also be prominent, although at wavelengths longer than a few metres it would be less bright than the radio galaxy in Cygnus. Radio waves would also be detectable from the moon and the planets; the planet Jupiter would be interesting to watch at long wavelengths, since it is characterised by irregular variations of radio intensity. Stranger even than Jupiter would be the steady flashing of the pulsars—most remarkable radio sources discovered in 1967 which emit short-duration pulses at regular intervals of about one second. But ordinary stars just would not be seen at all; the sun is a typical star and, if it were removed to the distance of the nearest stars, that is a few light years away, it is easy to calculate that the radio emission would not be detectable, even with the most powerful radio telescopes.

As soon as radio astronomy became possible, it was clear that it offered unparalleled opportunities for studying the shape of our own galaxy, for radio waves can penetrate with ease the dust lanes in the Milky Way which obscure most of the galaxy from optical telescopes. Much work has been carried out on this problem, and it is now known that our galaxy has a spiral structure rather similar to Andromeda. It has also been possible, using radiation at a wavelength of 21 centimetres, to find out how the galaxy is rotating. This radiation, which is emitted by hydrogen atoms, is one of the few atomic wavelengths which are easily detectable, and its importance lies in the fact that measurements of the Doppler shift (see p. 21) of the wavelength can be made. This means that it is possible to find the velocity of the hydrogen clouds relative to the solar system. One of the exciting new facts to emerge from these observations was that the central portions of the galaxy were expanding at speeds up to 50 kilometres/second. The reason for this motion is still not understood, and it

FIGURE 30 (a) Two reflectors of the Cambridge radio telescope. The most distant reflector is mounted on rails.

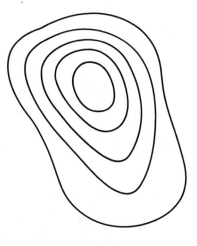

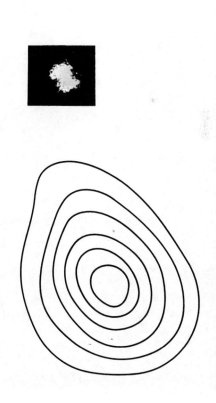

FIGURE 30 (b) A map showing the distribution of radio energy emitted from the galaxy Cygnus A. The radio signals come from two regions on either side of the visible galaxy. The lines are contours of radio intensity.

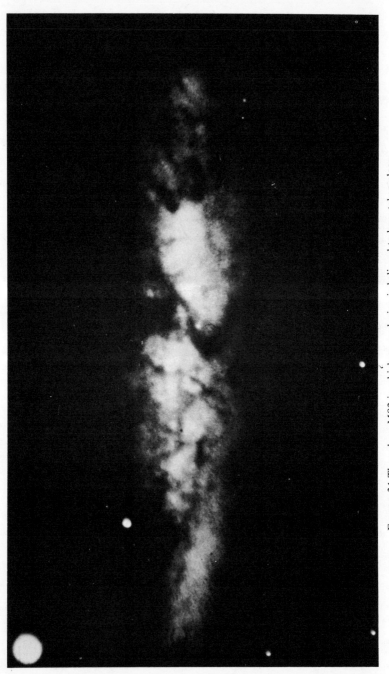

FIGURE 31 The galaxy M82 in which an explosion is believed to have taken place.

may be that an expansion phase is a normal event in the life cycle of most galaxies.

Most of the radio emission from the galaxy comes not from atomic hydrogen but from electrons moving at speeds near the velocity of light. Weak magnetic fields, perhaps ten thousand times less than the earth's magnetic field, are believed to thread all galaxies and, in the presence of such fields, electrons will be deflected into spiral orbits. This bending of the electron paths is accompanied by radio emission over a broad band of wavelengths known as synchrotron radiation, and most of the radio waves collected by radio telescopes originate in this way. Synchrotron radiation is similar to that from an excited atom (see Chapter 2, p. 29) but, since the electrons do not move in exact circles, the radiation covers a range of wavelengths; another difference is that the electrons have far greater energy than in an atom. The electrons themselves can be detected by measurements in space beyond the earth's atmosphere, and they form one component of cosmic rays, the name given to the high-speed particles of various kinds which continually scour the solar system and the whole galaxy. No one yet knows where the cosmic rays are generated, but they may be associated with galactic explosions, as we shall see presently.

Radio Galaxies

The radio emission from cosmic-ray electrons in our own galaxy is very weak in comparison with that from the peculiar galaxy in Cygnus. Only the nearest ordinary galaxies give detectable radio emission, and the radio galaxies are mostly far more distant than this. After the identification of the Cygnus galaxy, great effort was expended in intensive searches with optical telescopes of the positions of other radio galaxies to see if light could be detected from them. Very accurate pinpointing of the radio positions is necessary in this work, since photographs taken with large optical telescopes show such a wealth of stars and galaxies that it is difficult to be certain just which of them is also emitting radio waves. In the last ten years over 100 radio galaxies have been identified with certainty, and more than one-half of these appear to be large elliptical galaxies showing no spiral structure. However, most of the elliptical galaxies are *not* radio emitters, and it is still a mystery why a small number of these give such strong radio emission.

In general, the identified radio galaxies are at such a great distance that optical telescopes give only a very poor picture of what is taking place. Occasionally though, it is obvious that the galaxy is passing through a period of violent activity. Figure 31 is a photograph of the galaxy M82 in which an explosion appears to have taken place. The ragged filaments are known (from optical Doppler shift) to be expanding outwards at speeds up to 1,000 kilometres/second, which is five to ten times faster than galactic rotation speeds and is strong evidence of an explosive origin. Frequently the picture of a radio galaxy obtained with a radio telescope is markedly different from optical photographs, and a good example of this is the galaxy in Cygnus which was mentioned earlier. In Figure 30 (b) this radio galaxy is shown as mapped by the 'one-mile' radio telescope at Cambridge, and the emission contours indicate that the radio waves originate in two areas far beyond the limits of the optical galaxy. This galaxy is much more distant than M82, and little detail can be seen on optical photographs, but the greater extent of the radio source indicates that two clouds of matter have been expelled from the parent galaxy. It is natural to imagine that these clouds were blown out following an explosion of the kind that can be seen to be taking place in M82. Radio galaxies possessing two components like Cygnus are fairly common, amounting to about 70 per cent of the total, and it may well be that ordinary galaxies sometimes pass through an explosive phase during which they also appear as radio galaxies. If this period of activity is short compared to the entire life-span of a galaxy it would explain why radio galaxies are very rare compared to galaxies which do not emit radio waves.

While giving a brief description of radio galaxies, some mention must be made of quasars, since their discovery has stimulated enormous interest in the problem of galactic evolution. In the early 1960s a small number of radio sources had been identified with objects which, when observed with optical telescopes, looked exactly like stars. But the optical spectra of these sources were most unusual and contained atomic emission lines which were extraordinarily difficult to account for. The mystery was solved by Schmidt, in Pasadena, who found that the spectra fitted the familiar pattern for hydrogen if all the lines were red-shifted to give an increase of wavelength by a factor of $1 \cdot 158$. Other star-like objects which emitted radio waves were soon found to show even larger red-shifts once the secret for interpreting the spectra was known. Now the red-shift gives an immediate measure of distance (as will be explained shortly) if it is a 'cosmological' effect arising from the expansion of the universe. While there is still some uncertainty on this point, it seems likely that this is the simplest explanation of the red-shift. If it is true, then quasars are very distant objects, and the strength of the light waves from them can be explained only if they generate about 100 times more light than ordinary galaxies. In addition to this prodigious energy, the light intensity of some quasars has been found to fluctuate, which implies that they are incredibly compact systems measuring only a few light years across. When it is remembered that a typical galaxy has a size of about 100,000 light years, the peculiarity of quasars is even more evident. Many interesting theories have been propounded to account for quasars, and one suggestion is that a stellar explosion might trigger off a chain of similar explosions within a small region of sufficient stellar density. Another idea put forward by Hoyle and Fowler is that the 'gravitational collapse' of a massive gas cloud might also provide a source of sufficient energy; the treatment of this case involves general relativity, and a full analysis has yet to be worked out. It will probably be some years before the mystery of the quasars is clarified, but there is no doubt that their discovery, which came about through the advent of radio astronomy, has been one of the most exciting events in astrophysics.

Pulsars

As if the discovery of quasars were not sufficient radio astronomy turned up yet another mystery when pulsars were revealed at Cambridge in 1967. The remarkable feature about these sources, which lie at relatively small distances in our own galaxy, is that all the radiation which they emit is in the form of regularly spaced pulses. The pulses last, typically, for a few hundredths of a second, and are repeated with the accuracy of a clock.

At the time of writing the slowest pulsar has a period of 3.7 seconds and the fastest a period of 33 milliseconds. It has been found that the periods of several pulsars are slowly increasing, and this factor, among others, has led to the theory that pulsars are rapidly spinning stars which emit a narrow beam of radiation that sweeps around the sky to give a lighthouse effect. The increase of period then corresponds to the star slowing down as it loses energy. The interesting point about this theory is that only neutron stars can spin rapidly enough to produce the observed pulsar periods without flying to pieces. Neutron stars are very tiny stars, a mere 20 kilometres in diameter, yet they contain as much matter as a common star such as the Sun. The matter in them attains a density of over 1 million tons per cubic centimetre and consists mainly of neutrons—fundamental particles found inside the nuclei of atoms. Stars composed of matter in this state were predicted on theoretical grounds, but none had been found until the pulsar discovery. Neutron stars are believed to be formed as a result of the gravitational contraction which takes place when the burning phase of stellar evolution is completed. Gravitational collapse of the interior of stars was thought to lead to stellar explosions, like the famous Crab Nebula, which is now visible in the position of a star seen to erupt by Chinese astronomers in A.D. 1054. The discovery of a pulsar right at the centre of the Crab Nebula has done much to confirm the neutron-star hypothesis.

It is by no means clear how pulsars generate radio emission which is observed, but neutron stars must contain magnetic fields far more intense than those in common stars. One suggestion is that such rotating fields may accelerate electrons, in the space surrounding the star, to very great speeds, and they may subsequently convert their energy into electromagnetic waves.

The Origin of the Universe

It is perhaps an impertinence to suppose that the origin of the Universe will ever be fully understood. Nevertheless, there are certain basic questions to be asked which, even in our present state of ignorance, stand some chance of obtaining an answer. Cosmology is the name given to studies of this kind and, until recently, cosmology was a field in which mathematicians exercised their skill with little danger that direct observations might prove their hypotheses to be incorrect. Within the last decade, however, this situation has radically changed, largely owing to the new vistas revealed by radio telescopes.

When we look out into space using a powerful optical telescope the scene is dominated by the countless stars of the Milky Way, but beyond these lie other galaxies, becoming fainter and less distinct, as far as instruments can penetrate. The pioneering studies of these galaxies, now largely associated with the name of Hubble, took place around 1920–30, and one of the most dramatic discoveries was that of the red-shift. All but the very nearest galaxies, known as the local group, were found to be progressively reddened with increasing distance. This phenomenon is now generally believed to arise from the Doppler effect (see p. 21), and it can be explained if all the galaxies are moving away from the local group with velocities which increase in direct proportion to distance. At first sight this may seem unreasonable, but a little reflection shows that the situation would appear exactly like this if the whole Universe were expanding at the same rate. In this event every galaxy would be increasing its distance from its neighbours, and no matter from which galaxy observations were made, the others would seem to be receding from it.

Now the magnitude of the red-shift enables the velocity of recession to be calculated and, from this, one can estimate how close together the galaxies were at times in the past. It turns out that the galaxies must have been highly compressed some 10,000 million years ago, and this establishes, in one sense, an age for the Universe. It is interesting that the age of the earth, as measured from the radioactive decay of elements found in rock, is roughly one-half of this time span. Theories which suppose that the Universe has evolved by expansion from some highly condensed primeval state have been popularly referred to as 'big-bang' theories, since the initiation of the expansion was probably akin to some cosmic explosion. The galaxies as we now see them would, in this case, correspond to debris still flying away from the initial 'fireball'. If the expansion continues, the galaxies must eventually recede so far that they pass from view, and this might correspond to the end of the Universe.

The notion of a beginning and an end of the Universe has been regarded by some cosmologists as undesirable; such an idea, they hold, is equivalent to demanding that we exist at some special point in time. In a sense this may be compared with the medieval concept that the earth is central, the sun and the stars revolving about it. We now know that there is nothing special about our position in space, so why should not the same thing be true for time as well? By an extremely elegant theory evolved by Bondi, Hoyle and Gold, it was shown how the Universe could be so ordered that it always displayed the same overall appearance, despite the fact that it was expanding. For this to happen matter must be created all the time, so that new galaxies are formed at just the same rate as old galaxies are passing from view owing to the universal expansion. This theory was called the Steady-state theory, or the theory of Continuous Creation, and it removed the need to consider a beginning or an end in time. On the Steady-state theory the Universe has *always* looked roughly the same, although the individual galaxies are changing.

Since the two schools of thought about the history of the Universe are divergent, it is most important to see if observations can throw any light on the problem. Clearly all that is needed to decide the issue is to observe whether the galaxies really are thinning out in space, as demanded by the 'big-bang' theory, or not. The trouble is that any visible changes, on a human time-scale, are bound to be infinitesimal. However, because light takes a definite time to cross space, there is a means of looking into the past which offers great possibilities. The travel time of light from the sun is about 9 minutes; this means that when we see the sun *now* we are actually witnessing the surface of the sun as it *was* 9 minutes earlier. If the sun suddenly disappeared, 9 minutes would elapse before we saw it vanish. Similarly, we see the nearest stars as they were a few years ago, and the Cygnus galaxy as it was 600 million years ago. Light from the most distant galaxies which can be discovered using the most powerful optical telescopes has a travel time of one or two thousand million years, and it is frustrating that this time span is not quite sufficient for definite changes in the galactic density to be noticeable. One of the difficulties of using light waves for this analysis is that the red-shift displaces most of the visible light into the infra-red portion of the spectrum where measurements are difficult to make.

Now it is a lucky chance that the radio window in the atmosphere permits us to collect radiation from distances much greater than can be reached by means of light waves. One radio galaxy, known as 3C295, has been identified with a galaxy at a distance of about 5,000 million light years, so that, in this particular case, we know that we can reach back 5,000 million years in history. This galaxy is a strong radio emitter, and radio waves one hundred times fainter can be picked up from galaxies which are believed to be much farther off still. Thus, with radio telescopes at any rate, we really do seem to be observing the most distant galaxies at a time close to the origin of the Universe on the 'big-bang' cosmology. Unfortunately no direct measure of red-shift is possible using radio waves, since insufficient energy is radiated in the form of definite atomic wavelengths. However, it is possible to determine the average radio emission properties of radio galaxies fairly close to us and, from this, to gain some idea of the distance of radio galaxies which appear much fainter on the records.

One method of investigating whether radio galaxies really were closer together in the past is simply to count the number which radiate more strongly than some chosen level. As the chosen strength, or intensity, is made smaller and smaller, the number of galaxies increases, since there are always more weak ones to be seen than intense ones. It is not difficult to calculate how the number should increase provided that space is, on average, uniformly filled with galaxies. The counts can then be compared with the theoretical estimates to see if the assumption of a uniform population is correct or not. Much effort has been expended on simple observations of this kind, the main emphasis being to make reliable counts of distant radio galaxies which are the ones most likely to show cosmological effects. Some observations made at Cambridge are shown in Figure 32, from which it may be seen that the fainter radio galaxies are much more numerous than would be expected on the basis of the steady-static cosmology. While ingenious attempts have been made to retain the basic concepts of the steady-state theory and yet to explain the counts of radio galaxies, the evidence is strongly in favour of a Universe which evolves with time.

Once evolution is admitted, other effects besides the steady thinning-out of galaxies may be expected. The galaxies themselves have some 'life-history' and, since the distant regions of space appear younger because of

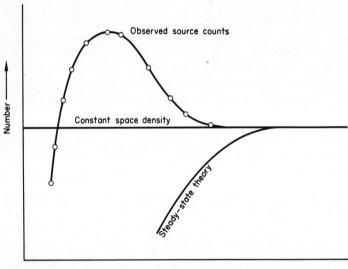

FIGURE 32 Observed counts of radio galaxies showing a marked departure from values predicted by the steady-state theory.

the light travel-time which we have previously discussed, it is natural that the galaxies themselves may look different, in addition to being more closely packed. Now the period of time during which galaxies become unstable and blow off clouds of matter to become radio galaxies is likely to be quite a small fraction of the total life of a galaxy. So, on an evolutionary cosmology in which most of the galaxies were formed at roughly the same time, we may expect to find far more radio galaxies at one time than at another. From the observations that have been made during the past few years the evidence points to the fact that the radio-galaxy phase occurs when galaxies are quite young. While far more data is needed before the evolutionary cycle is satisfactorily established, the important fact is that cosmology is now an observational science, and it seems to be definitely established that the Universe came into being at some singular point in time.

Quite recently the radio window in the atmosphere has revealed evidence for a 'big-bang' Universe in an entirely different way. When all the matter of the Universe is highly compressed, at the very beginning of the expansion, a large fraction of the energy of the system is in the form of electromagnetic radiation of very short wavelength. The situation resembles the state of affairs deep inside the sun, where the temperature is extremely high and where there is a high intensity of γ-rays. As the expansion of this 'primeval fireball' proceeds, calculations show that the γ-rays become converted to longer and longer wavelengths. For a Universe of the size that we now see, it may be calculated that the radiation would have been transformed from γ-radiation to infra-red radiation. The spread of wavelength would, however, be considerable, and there should still be measurable quantities left in the microwave band of a few centimetres wavelength. It has recently been found that the earth is, indeed, bathed in such radiation, and this provides very convincing evidence that a 'big-bang' actually took place about 10,000 million years ago.

FIVE

SEEING THE VERY SMALL: X-RAY DIFFRACTION

X-rays in General

X-rays, discovered by a German physicist, Wilhelm Konrad Röntgen, rather more than seventy years ago, first captured the public imagination because of their unusual property of being able to pass through materials, such as black paper, cardboard or human flesh, that would stop ordinary light. X-rays were, however, absorbed by lead or even bone; hence it was possible to obtain shadow-graphs of the skeleton inside the human body and thus to observe breaks in bones, or foreign metallic bodies, such as pins or coins, that had been swallowed. There are people still alive, seventy-five years old and upwards, who were among the first to be 'X-rayed' for broken bones or (less successfully) to diagnose the presence of 'stones' (urinary calculi) in the kidney or bladder.

Soon it was found that the almost random, happy-go-lucky use of X-rays that had followed Röntgen's discovery had to be restricted because, in some cases, people developed serious and very unpleasant skin burns as the result of X-ray exposure. This destructive feature of X-rays has its uses, however. Cancerous cells were later found to be more susceptible to destruction by X-rays than normal cells, and therefore, by using the right kind of X-rays and by carefully controlling exposure times, cancerous cells can be destroyed without also destroying the patient.

It was found that X-rays themselves had a range of transmissibility. 'Soft' X-rays were relatively easily stopped, whereas 'hard' X-rays were much more penetrating. Both affected photographic plates so that their impact could be recorded; both would make gases electrically conducting (by knocking electrons out of the atoms and thus ionising them), which provided another way of detecting them and of measuring their intensity. Controversy raged as to the nature of X-rays. Röntgen himself was inclined to think that they must be some kind of *longitudinal* wave similar to those of sound but of much shorter wavelength. Others, because of their ionising properties, believed them to be some kind of very penetrating fast *particle*. But, as is now well known, X-rays are in fact *transverse electromagnetic waves*, similar to radio waves but of very much shorter wavelength. The X-rays used in hospitals are mostly hard X-rays, very penetrating and about 100,000 times (or more) shorter than light waves. But, although soft X-rays, which have longer wavelengths, are easily absorbed and therefore less suitable for radiography, they are the most suitable for use by the crystallographer, who studies the arrangement of atoms in crystalline solids.

The Structure of Matter

Before we can explain why soft X-rays are the most suitable for this purpose, however, we must say something about the arrangements referred to. Most people would think that they know the difference between a solid, a liquid and a gas. Taking the last first, a gas expands to fill whatever vessel, of whatever shape and size, is used to contain it. If released from any vessel it expands indefinitely within the earth's atmosphere. Some male moths can locate a female of their own species from over a mile away, because special molecules emitted by the female travel so widely that they can be 'scented' over these large distances. Liquids, on the other hand, are limited in volume at any given temperature, but are able to take up the shape of the vessel into which they are put. If they are very viscous this may take time, but the viscosity may be lowered by raising the temperature; warming treacle will make it pour more easily. A solid has both a definite volume and a definite shape; it is rigid. Sir Oliver Lodge, one of the great physicists of the early part of the twentieth century, once wrote that, 'The most extraordinary thing about a poker is the fact that, if you lift it up by one end, the other end comes up too.' Of course there are borderline cases: hair and muscle are not rigid, paper can be folded.

However, the complex nature of ordinary solid matter is not satisfactorily explained by a simple reference to rigidity. Glass is rigid at ordinary temperatures but, if it is heated sufficiently it will flow; it becomes liquid gradually. Plastics have acquired their name because, if warmed, many of them change their shape, that is, they become plastic. These are what are usually called *amorphous* solids. They have no definite melting-point. Solids like ice, which melts into water at a temperature so well known that it is used as a fixed point on our temperature scales (0° C or 32° F), are *crystalline* solids. We now know, mainly through the techniques of X-ray diffraction that will be described in this chapter, that crystalline solids consist of a regular array of atoms (which may be ionised) or of groups of atoms or molecules (see Figures 33, 34, 35). In many metals, such as aluminium, copper or gold, the arrangement is a simple one in which atoms of the metal, say copper, are packed as closely as spheres can be packed, each sphere being surrounded in space by twelve similar spheres. In the virus of poliomyelitis (and some others) there is a somewhat similar arrangement (Figure 36), but this time of molecules, each of which in itself is a very complicated arrangement of atoms but which happen to have an effectively spherical molecular shape.

Most crystalline solids, however, are not close-packed, even though they may be thought of, to a first approximation, as consisting of spherical atoms of a definite radius. If the solid is one like rock salt (sodium chloride), quartz (silicon dioxide) or calcite (calcium carbonate) (Figure 34), which each contain more than one kind of atom of different radii, then a simple close-packed arrangement is not possible. Even some elements such as carbon have structures which are far from close-packed because, in

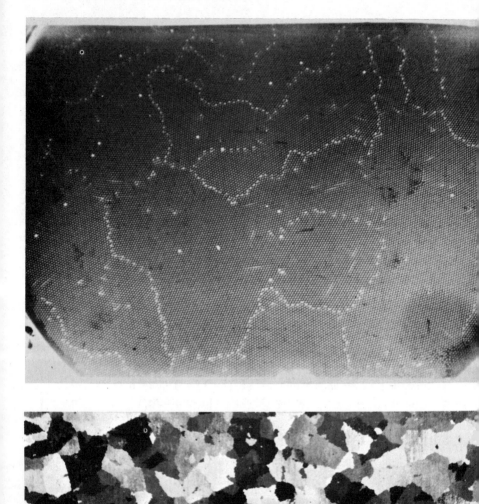

FIGURE 33 Conglomerate of crystals in a sheet of aluminium (in ordinary sheet metal these would be only about 10^{-4} cm across), and the corresponding arrangement of atoms (simulated by bubbles. After W. L. Bragg and J. F. Nye.)

these cases, the atom itself exerts forces which are directed in space so that the number of its neighbours is limited. In diamond each carbon atom has four neighbours only, tetrahedrally arranged about it, and the density is much less than if it were close-packed.

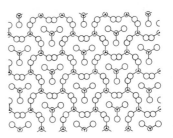

In more complicated solid materials, whether minerals, chemicals or biological substances, the arrangement is often a very intricate one, but all crystalline solids are characterised by periodicity and symmetry. Periodicity means that a basic pattern is repeated in space exactly similarly at regular intervals (at least, it is *ideally* so repeated in practice—the pattern may exhibit mistakes like the faults or unevennesses in hand-knitted or hand-woven material). Symmetry, one example of which is shown in Figure 34, is a feature of the basic pattern itself. In some cases, for example in crystalline minerals such as

FIGURE 34 Diagrammatic representation of the arrangement of atoms in calcite (calcium carbonate, as found in seashells, etc.) looking along a three-fold symmetry axis of the crystal.

quartz, or in artificially grown crystals such as sugar, the external crystal faces reveal something of the internal symmetry. But a diamond is crystalline whether it has grown as an octahedron or as a mis-shapen

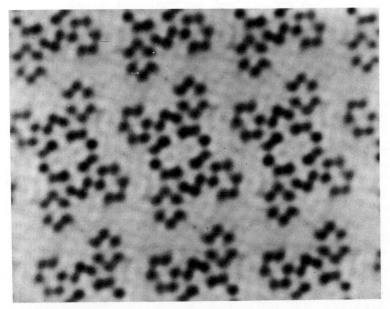

FIGURE 35 Electron-density representation of the molecules of metal-free phthalocyanine as projected along a prominent crystal axis.

lump—the internal arrangement is basically the same. Many chemicals consist of fine powders, each grain of which is a tiny 'single crystal' or a conglomerate of 'crystallites'. Most metals, whether in the form of wire, sheet or massive material, are conglomerates (Figure 33): each irregular bit of the conglomerate consists of many millions of atoms, regularly arranged; but the bits are not parallel.

On the other hand, an amorphous solid is one in which there is no basic unit of pattern regularly repeated. In glass, for example, there are silicon and oxygen atoms and some metal atoms, such as sodium, potassium, iron, etc.; but, although atomic forces exist and although the atoms behave as if

FIGURE 36 Diagram of a crystal of poliomyelitis virus. (After Dr. C. Bunn, *Crystals, their Role in Nature and in Science.*)

they have a definite 'size', so that each silicon atom has four oxygen atoms as nearest neighbours and each oxygen atom links two silicon atoms, yet there is no symmetry and no periodicity. When very old glass 'devitrifies' it crystallises, but normally glass is like a liquid in its lack of periodic pattern. The only difference between a liquid and an amorphous solid is one of movement of the individual atoms within the volume occupied. In liquids the atoms constantly roll over one another, changing their positions. In all solids, whether crystalline or amorphous, the atoms are vibrating about fixed positions. The amplitudes of vibration and, to a certain extent, their frequencies increase as the temperature rises. At some fixed temperature the vibrations become so important relative to the periodicity that the regularity of arrangement in a crystalline solid breaks down and it either melts or changes over to some other periodic arrangement which is more stable at the higher temperature (a phase transition). In an amorphous solid the effect of the rise of temperature is much more gradual. The solid first changes to a very viscous liquid and the atoms, which were formerly vibrating about fixed but non-periodic positions, begin to roll over one another or to change places. As the temperature rises, the substance becomes a true liquid without any abrupt change in the viscosity curve and without any absorption of energy (latent heat) at one definite 'melting' point.

Early philosophers were greatly intrigued by the smooth shining faces on crystals, faces which might be of relatively different shapes and sizes in different specimens of the same substance but which were inclined to one another at certain definite angles. This 'Law of Constancy of Inter-facial Angles' led them to believe that such crystals were built up of regularly arranged units of pattern and, although they were not sure what the 'unit' might be, because they were not then as familiar as we are with the idea of atoms and molecules, yet some shrewd guesses were made. These guesses for simple substances such as rock salt, combined with a knowledge of the density (mass in grammes per unit volume in cubic centimetres) led to the deduction that the average minimum distance apart of the atoms in such solids must be of the order of 1–3 Å (Å stands for the Angström unit, a distance of 10^{-8} cm.).

X-ray Diffraction

The wavelength of ordinary light waves is of the order of several thousand Å (or $\frac{1}{1000}$ cm), and so it is quite hopeless to expect to see, or to photograph, individual atoms with them. With the modern electron microscope it *is* possible to photograph a virus particle and to 'see' regularly arranged individual virus molecules on such photographs (Figure 37), because these molecules are very large, each over 300 Å across. Even so, we are not seeing the molecule directly, but only the photographic image of it obtained (in the electron microscope) using electrons and not light waves; we do, of course, look at the photograph by means of light waves.

Röntgen, whose investigation of the X-rays that he discovered was amazingly thorough, had found that, if he shone a beam of X-rays through a small hole the X-rays did not seem to bend slightly round the edge of the hole, as ordinary light would do. This phenomenon in light is due to diffraction, which has been explained in Chapter One. The absence of this kind of diffraction led Röntgen to the conclusion that his newly discovered rays had a wavelength of only about 1 Å, but he knew of no way of measuring this sort of length exactly. He did, in fact, shine his X-rays through powdered and single crystals, and it was perhaps bad luck that he did not make the observations that were made, also in Germany, some seventeen years later.

'The experiments on the permeability (for X-rays) of plates of constant thickness cut from the same crystal in different orientations, which were mentioned in my first Communication, have been continued. Plates were cut from calcite, quartz, turmaline, beryl, aragonite, apatite and barytes. Again no influence of the orientation on the transparency could be found.'

'Ever since I began working on X-rays, I have repeatedly sought to obtain diffraction with these rays; several times, using narrow slits, I observed phenomena which looked very much like diffraction. But in each case a change of experimental conditions, undertaken for testing the correctness of the explanation, failed to confirm it, and in many cases I was able directly to show that the phenomena had arisen in an entirely different way than by diffraction. I have not succeeded to

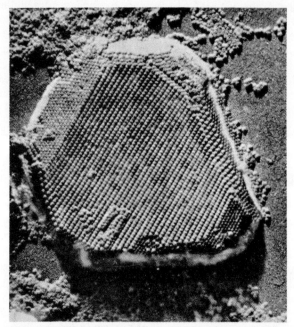

FIGURE 37 Electron microscope photograph of a southern bean mosaic virus crystal. (From Labaw and Wyckoff. Arch. Biochem. Biophys., **67**, 225 (1957), Copyright Academic Press.)

register a single experiment from which I could gain the conviction of the existence of diffraction of X-rays with a certainty which satisfies me.' (Translated and quoted on p. 5 of *Fifty Years of X-ray Diffraction*, see later reference.)

It happened in 1912 that a student named Paul Ewald had made a theoretical investigation of the optical diffraction effects to be expected from an anisotropic three-dimensional array of scattering particles such as a crystal. He consulted a young lecturer, Max von Laue, about some of his results. This conversation brought home to von Laue for the first time the fact that crystals are such three-dimensional arrays, and he realised that, if the average distance apart of the atoms in a crystal is about 1–3 Å, and if this is also the wavelength of X-rays, then X-rays might give an observable diffraction pattern when passed through a crystal. He was not quite sure. It was known that the atoms in a crystal must be vibrating. Perhaps the effect of the vibrations might be such that no diffraction pattern could be clearly seen? This is not in fact the case. For example, in diamond (one of the substances studied) the average amplitude of thermal vibration at room temperatures is only about 0.05 Å, as compared with the nearest carbon-to-carbon distance of 1.54 Å. In softer compounds the vibrations are larger, but the substance usually melts or changes its structure when the amplitude of vibration becomes as much as $\frac{1}{5}$ of the atomic separation. However, at the time it was thought that the vibrations were much larger than this. The whole story is described very dramatically in *Fifty Years of X-ray Diffraction*, edited by P. P. Ewald and published by the International Union of Crystallography (obtainable only from the publisher N. V. A. Oosthoek, Utrecht, The Netherlands).

Von Laue was not himself an experimentalist, but was a theoretical physicist. So he asked two young students to try the experiment of placing a crystal in a narrow beam of X-rays to see whether there could be recorded on a photographic film any diffraction pattern at all. Using copper sulphate, zinc blende and diamond, they did observe such patterns, but von Laue was not really able to interpret them in terms of the atomic arrangements which had caused them. The principal reason for this failure was that they were using X-rays having a range of wavelengths, and that there was at the time no real clue as to what atomic arrangements occur in the particular kinds of crystal they used. The history of the reasoning and the experiments which led W. H. and W. L. Bragg to the correct analysis of many simple crystal structures (the first of which, by W. L. Bragg, was zinc blende), studies for which they were jointly awarded a Nobel Prize for Physics in 1915, has been related in detail by them in *X-rays and Crystal Structure*, Vol. I, to which reference may be made.

Briefly, however, the argument went like this. Rock salt probably has a chessboard-like structure, sodium and chlorine occurring alternately (like the black and white squares), but in three dimensions. A model (not known to W. L. Bragg at the time) showing such a structure was actually constructed by Wollaston (1766–1828), and is in the Science Museum, London. Another model, constructed by Dr. Crum Brown out of knitting-needles and balls of wool, is in Edinburgh University. The distance apart of the atoms in such a structure can be exactly calculated (without the use of X-rays) as follows. The space occupied by the two unlike atoms, sodium and chlorine, will be just twice the cube of the distance between them: $2d^3$, and the weight of material in that volume will be $2Dd^3$, where D is the density of rock salt. But, since the atomic weights of sodium and chlorine are 23.0 and 35.6 respectively and the unit of atomic weight is $1.66 \cdot 10^{-24}$ g (the reciprocal of the Avogadro number, which is the number of atoms in one gramme-atom, known to be constant), this joint weight of sodium + chlorine must be $(23.0 + 35.6) \times 1.66 \cdot 10^{-24}$ g. We know, by direct measurement, that $D = 2 \cdot 163$ g/cm³, and hence $d = 2 \cdot 814 \cdot 10^{-8}$ cm.

The general formula for a unit of repeat of any possible shape (which must be one which, when repeated three-dimensionally, fills space) is

$$V_c \cdot D = nM/N$$

where V_c is the volume in cm³ of the 'unit cell';

D is the density in g/cm³;

n is the number of molecules each of molecular weight M contained in V_c;

N is the Avogadro number $6 \cdot 023 \cdot 10^{23}$.

W. L. Bragg was able to confirm that rock salt does indeed have the chessboard pattern and so does sylvine, which is potassium chloride. He

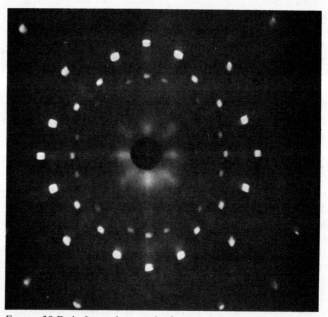

FIGURE 38 Early Laue photograph of rock salt, taken with the X-rays incident along a cube (fourfold symmetry) axis. Radiation from a silver target.

did this by consideration of the positions and intensities of the spots on the 'Laue' photograph (Figure 38), and of the changes in intensity which should and do occur when sodium is replaced by potassium.

The Interpretation of X-ray Patterns

In general, the use of an X-ray beam containing a range of wavelengths complicates the problem of interpretation, although the fact of wave diffraction by a periodic system of scattering particles is not difficult to demonstrate. A simple experiment with a handkerchief or, better still, with a piece of fine nylon fabric will illustrate it. In this case, however, one must use ordinary light, because the holes in the fabric which are going to diffract the waves are usually over $\frac{1}{100}$ cm apart, or a million times greater than interatomic distances.

If a point source of bright light, such as a distant lamp, is looked at through a piece of fabric, what is seen is a pattern of points of light

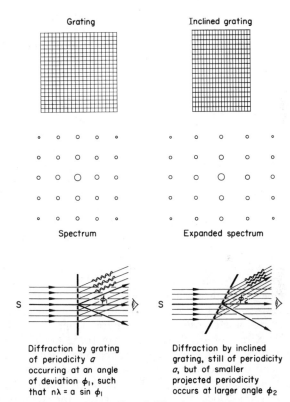

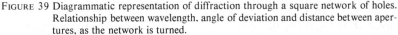

FIGURE 39 Diagrammatic representation of diffraction through a square network of holes. Relationship between wavelength, angle of deviation and distance between apertures, as the network is turned.

arranged on a square mesh. The fact that this is a *diffraction* pattern is proved by inclining the fabric so that the threads appear to come *closer together*. The points of light then appear to move farther apart (Figure 39). This is the consequence of the diffraction equation, which relates the distance *a* between the holes (scattering centres) to the angle of diffraction ϕ and to the wavelength λ of the incident and scattered waves. A bright point of light occurs in the image when the difference of path-lengths of the scattered waves from successive holes is an integral number of wavelengths (see p. 20). Expressing this mathematically gives $n\lambda = a \sin \phi$ so that $\sin \phi$ increases as a (or a projected) decreases, for any given integer n.

If the distant light is a white light the pattern will be coloured; it is then a number of square meshes superimposed, the red mesh (longest wavelength) being largest, the blue mesh (shortest wavelength) smallest. In other words, for a given a and n, $\sin \phi$ varies directly with λ. Each spot on the pattern is drawn out into a little spectrum pointing towards the centre of the multiple mesh. This illustrates only a two-dimensional case of diffraction, and of ordinary light, but the principle is the same if we use a three-dimensional diffraction grating, that is, a crystalline solid, with X-rays.

Coming back to our illustration: if the light is monochromatic (say from a sodium lamp), then only one coloured mesh of diffraction points is seen (a yellow one for sodium light) and, from the distance apart of the points on this diffraction pattern, which could be recorded photographically, it would be possible to calculate a if the wavelength λ were known. Now suppose that, instead of having a pure linen or cotton handkerchief, the fabric was one in which every tenth thread in one direction was made of wool or of nylon, or was a different thickness. Or suppose the mesh was not square. There would still be a diffraction pattern, but it would be a different one. It would be different again if any other *systematic* variation were made. In fact, any fabric, simple or complicated, with a regularly repeated pattern, would give a diffraction pattern for radiation of any wavelength; but a useful pattern would be obtained only if the wavelength of the radiation were of the same order as, or somewhat less than, the basic repeat distance in the fabric.

In crystals this basic repeat distance is of the order of 10 Å, because although the radii of the atoms are always less than about 2 Å, there are usually a number (though not a very large number except in biological substances such as viruses or other proteins) of atoms in each unit cell. What we need, to obtain really useful patterns from such crystals, are monochromatic X-rays of wavelength less than 10 Å. Actually the best X-rays to use are those having wavelengths from 0·5 to about 2·5 Å, because longer waves tend to be so heavily absorbed, even by very small crystals, that no useful diffraction pattern is obtained, while shorter wavelengths give diffraction spots which are too close together for comfortable measurement. Moreover, it is found that the intensity of diffraction is proportional to the cube of the wavelength, so that, apart from absorption, the longer the better. The problem, of course, is to deduce the shape and size of the repeat unit, the coordinates of the atomic centres within the repeat unit and the amplitudes of the atomic vibrations, all from measurements on the diffraction pattern observed. Since in practice this pattern may have many hundreds or even thousands of diffraction spots, whose positions, intensities and shapes can be measured, there is a mass of data available.

However, the first requirement is to obtain a monochromatic (or nearly monochromatic) beam of X-rays and to determine the wavelength accurately. This was first accomplished by W. H. Bragg, who designed an apparatus known as the ionisation spectrometer. He knew that, provided the voltage which operated his X-ray tube was greater than a certain minimum value, the material of the 'target' in the tube would emit not only 'white' or 'continuous' X-radiation having a range of wavelengths but also a spectrum of certain definite wavelengths characteristic of the target material. A copper target, for example, emits three useful X-ray lines, known as the $CuK\alpha_2$, α_1 and β, of wavelengths 1·544, 1·541 and 1·392 Å respectively and of relative intensities 5 : 10 : 2 at the target face. The optimum voltage at which to run a tube having a copper target is about 35,000 volts (35 kV), because at this voltage the characteristic CuK radiation is very intense relative to the white radiation. W. H. Bragg succeeded in measuring these wavelengths (and those from several other target materials) by using rock salt as a diffraction grating and by measuring diffraction angles on his ionisation spectrometer. W. L. Bragg had shown that such a crystal grating diffracts X-rays as if it were composed of equally spaced sheets of atoms actually reflecting the incident beam, but only at those special angles θ_n such that

$$n\lambda = 2d \sin \theta_n$$

where n, as before, is an integer, but d is now the interplanar spacing and θ_n is one-half of the angle of deviation from the main beam, for a given peak. This equation, which expresses *Bragg's Law*, is basic to X-ray-diffraction theory.

Since the structure of rock salt was known, the values of d corresponding to different sets of crystal planes could be calculated quite easily, and hence, by measuring θ_n, λ could be determined. Using λ, d could be determined for sets of atomic reflecting planes in any other crystal under investigation by measurement of the diffraction angles θ_n at which selective reflection occurred. Using the original ionisation ('Bragg') spectrometer, θ was measured by mounting the crystal, usually 2 or 3 mm across, at the centre of the spectrometer table with a prominent crystal axis vertical and rotating it slowly. The X-ray beam from a 'Coolidge tube' was well shielded to prevent unwanted radiation from reaching the experimenter, and the useful beam was collimated so that only a narrow beam of cross-section smaller than the crystal was actually incident upon it. As the crystal rotated, successive crystal planes would come to their respective diffraction angles θ (Figure 40), and the diffracted beam would flash out, to be collected by an ionisation chamber moving at twice the rate of the crystal. The gas in the ionisation chamber varied in electrical conductivity with the intensity of X-rays diffracted into it, and this variation was measured by means of a gold-leaf electroscope, giving a record of the type shown in Figure 41.

It is useful to eliminate the $K\beta$ peak from the characteristic spectrum, leaving only the close $K\alpha_1$ α_2 doublet, and this is easily done by inserting a filter into the primary beam. For CuK radiation a filter of thin nickel foil is suitable.

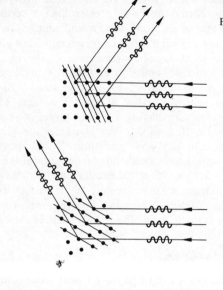

FIGURE 40 When a crystal is rotated in a monochromatic X-ray beam the various planes of atoms are able to reflect only at particular angles; the angle at which reflection takes place depends on the spacing of the planes. The more widely spaced planes give reflections near the primary beam; the more closely spaced planes reflect at the larger angles. (By courtesy of Dr. Chas. Bunn.)

FIGURE 41 Record of reflections from rock salt (NaCl) made with the Bragg ionisation spectrometer, using radiation from a palladium target.

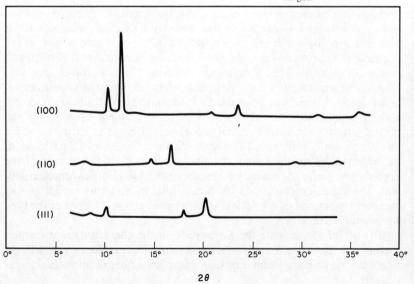

2θ

A very similar, but much more sophisticated arrangement, is used today in modern diffractometers, where the diffraction spectrum is either recorded on a chart or on tape or cards to be fed into a computer for automatic processing. The advantage of the spectrometer or diffractometer is that the position, spectral line profile (if required) and intensity can all be recorded simultaneously. The disadvantage is that measurements of diffraction from different sets of crystal planes must be made one at a time. If

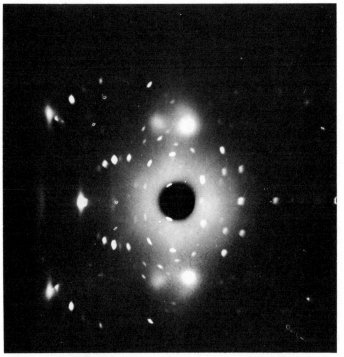

FIGURE 42 Laue photograph of sorbic acid, showing sharp spots due to white radiation diffracted by the 'average' structure and diffuse spots due to CuK radiation diffracted by thermal vibration of the molecules.

the crystal is quite stable this is no disadvantage, but if it is changing (for example, evaporating, dehydrating, rusting or undergoing any other kind of slow chemical reaction) the results of successive readings may not be comparable. Moreover, if one is interested not only in the positions of the atoms but also in their vibrations and in disorder effects which modify the ideal periodic arrangements, then it may be necessary to measure the intensity of diffraction between the main peaks. Diffuse streaks and patches (Figure 42) give the clues required to study these disorder effects. Such background scattering is much more readily recorded on photographic films or plates. These have the additional advantage of providing a simultaneous and permanent record of the diffraction from a large variety of crystal planes, but the disadvantage that measurement of intensity requires a separate operation. If the intensity over any area (and not just a peak

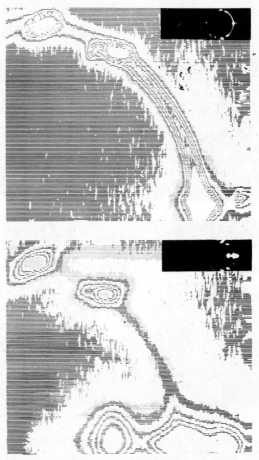

FIGURE 43 Isodensitracer record of intensity from stationary crystal photographs which show both a continuous ring and superimposed spots. Two stages of a chemical reaction in the solid state. It would be impossible to measure the ring intensity by orthodox photometric techniques or diffractometer.

measurement) is required, then the most promising instrument to use is the isodensitracer (Figure 43), which is also used by astronomers in measuring the light reaching us from distant stars and galaxies.

The actual process of diffraction from a three-dimensional array of repeated units has been well illustrated, on a two-dimensional scale, by a series of patterns prepared by Professor H. S. Lipson of Manchester. These are shown in Figure 44. The diffraction pattern produced by shining a narrow beam of light through a single circular hole, smaller than the beam, would be a circular patch surrounded by concentric rings of decreasing intensity. If we use an array of holes, imitating the arrangement of atoms in the 'unit' of repeat (which may be a symmetrical set of atoms or a molecule or group of molecules), then we get what is known as a *Fourier transform*

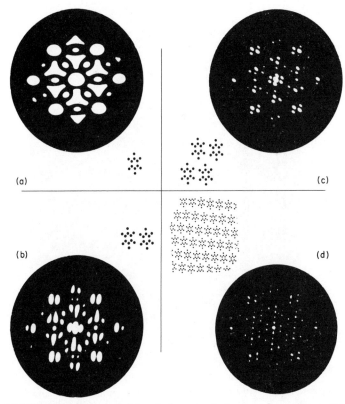

(a)

(c)

(b)

(d)

FIGURE 44 Diffraction patterns of: (a) a single molecule (molecular transform); (b) two parallel molecules; (c) four regularly-spaced molecules; (d) a net-work of similar molecules. The last diffraction pattern may be compared with Figure 59. (By courtesy of Prof. H. S. Lipson.)

of the 'unit'. This consists of a series of patches forming a pattern from which, by a mathematical analysis, the original arrangement can be deduced. Such a Fourier transform may be obtained with waves of any length and any object of an appropriate size. Radio waves would be diffracted by an aeroplane to give a Fourier transform of the aeroplane. X-rays are diffracted by single molecules to give a Fourier transform which is observable, however, only when millions of molecules are diffracting in parallel. A system of two or more such 'units', regularly repeated, will result in the addition of one or more systems of parallel interference fringes to the diffraction pattern, the 'spacing' of the fringes being inversely proportional to the repeat spacing of the rows of 'units'. If the repetition is carried, two-dimensionally, to infinity, then the original Fourier transform is crossed by an infinite series of sets of interference fringes which reduce it to a pattern of spots, arranged as a regular two-dimensional network and of varying intensity. From the positions and intensities of these spots it

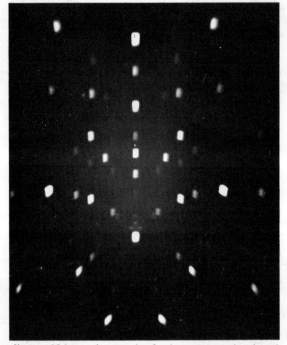

FIGURE 45 Laue photograph of a large rectangular-shaped diamond plate, taken on a plane X-ray film inclined so that the transmitted incident beam just misses it. (Whole radiation from a copper target.)

should be possible to deduce (by rather tedious mathematics) the positions of all the holes in the original *diffraction grating*, the two-dimensional repeat of the 'unit'. Thus we would find not only the original arrangement of holes in the unit (atoms in the molecule or in whatever grouping is basic to the crystal) but also the repeat periodicity in any direction (the actual size and shape of the *unit cell*).

There are certain features of such diffraction by an extended array of scattering centres which are not at all obvious until they are pointed out, and there are also in practice, limits to the extent to which the analogous X-ray-diffraction patterns may be observed, measured and interpreted. In the first place the original diffraction grating never is infinite. It would be quite impossible to make such an infinite series of holes in a sheet of metal or cardboard or plastic; and, of course, no crystal is infinite either. The crystals used in modern structural studies are usually less than 0·5 mm in any dimension and may only be one-tenth of this. If they are in powdered form they are a good deal smaller still. It would take 500,000 unit cells

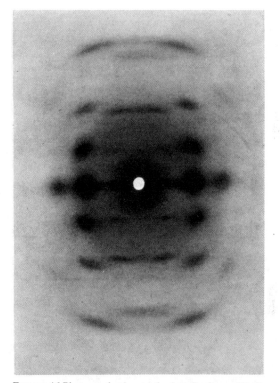

FIGURE 46 Photograph, due to the late Professor W. T. Astbury (courtesy of *Endeavour*), of natural silk, in which the component crystallites are small enough for a general spreading of all reflections. There is probably also some disalignment of crystallites contributing to the spot size, but disalignment alone would expand high-angle spots far more than low-angle ones (CuKα radiation).

each 10 Å across to measure 0·5 mm. Half a million is a lot, but it is not infinity. Theoretically, the larger the crystal, the smaller the diffraction spots in the diffraction pattern. In practice, it doesn't work out like this at all. Geometrically, when we take a photograph of a large crystal with a collimated X-ray beam (Figure 45) we get large spots on our photograph! If the crystal is stationary each spot is simply a distorted image of the crystal cross-section, modified by the detail of the X-ray source, by any crystal inhomogeneities and by the angle at which the diffracted beam hits the photographic film. It is also true, however, that, if we use *very* small crystals (10^{-5} cm or less across), then our diffracted beams begin to spread, as the theory would predict (Figure 46). The ideal size, for practical purposes, of our crystal (or of our diffraction grating, whatever it may be) is intermediate between these two extremes.

In the second place, if our elementary holes (or atoms) each acted as a stationary *point* source of scattered light (or of X-rays), then the diffraction patterns would stretch to infinity with the same *average* intensity. In practice, this is not so, whether we are dealing with holes drilled in a sheet of metal, with the gaps between the threads of a piece of fabric or with the atoms in a crystal. Each scattering centre which makes up the 'unit' is itself extended in space—it is not a point source. Different parts of a finite source will interfere with one another, and the result of this interference is that the diffraction pattern gradually decreases in average intensity from the centre outwards; thus the diffraction pattern of a distant lamp that we see through a handkerchief is limited in size, fading away to the outside. Similarly, the pattern of spots in Figure 44, which corresponds with the extended network of molecules, is finite in extent. In the case of atoms, X-ray scattering is by the electron cloud, and is greater, the heavier the atom. The amplitude of the wave scattered by a 'free' electron is known, theoretically, to be e^2/mc^2, where e, m are the charge and mass of the electron, and c is the speed of light. The forward scattering of an atom of atomic weight Z (that is, having Z extra-nuclear electrons) is Ze^2/mc^2. However, the scattered waves from different parts of the electron cloud of a single atom interfere with one another, and the path differences increase as the angle of deviation from the forward direction increases. Figure 47 shows a series of *atomic scattering curves* for different kinds of atoms, or f (short for 'form') factors, as they are often called. The vibration of the atoms increases their effective spread in space and causes an even greater falling of the value of f as θ increases or as λ decreases so that, at higher temperatures, the whole diffraction pattern from a crystal is somewhat weakened,

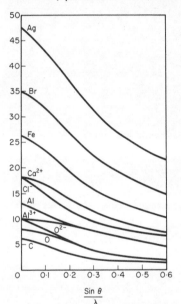

FIGURE 47 Graph showing basic atomic scattering curves for a number of atoms and ions of different scattering powers.
Thermal vibration will reduce the scattering power at high angles still further.

and at low temperatures it is enhanced. On the other hand, the diffuse patches and streaks previously referred to, which are a direct result of the thermal vibration, are enhanced at high temperatures and weakened at low ones (Figure 48).

The third feature is one that affects the actual measurement of X-ray-diffraction patterns. It is a consequence of the Bragg law, $n\lambda = 2d\sin\theta$. The maximum value of $\sin\theta$ is 1; for which $\theta = 90°$, and 2θ, the angle of deviation of the X-ray beam, is $180°$. This means that, even if there were no falling off of intensity with increasing θ, it would still not be possible to observe diffraction effects beyond a certain limit. What that limit is for any given crystal depends upon the X-ray wavelength used, since $\sin\theta = n\lambda/2d$. Let us consider this relationship a little more in detail, taking the simplest possible model of a monatomic cubic crystal. The atoms in this or any other crystal can be pictured as lying in equally spaced parallel planes.

(a)   (b)

FIGURE 48 Diffraction photographs of a stationary organic crystal (penta-erythritol) taken with the whole X-radiation from a copper target: (a) at room temperature; (b) at a low temperature. The sharp Laue spots are increased in number and intensity at the low temperature, the diffuse thermal spots are weakened.

The interplanar spacing d is different in different directions (Figure 49), but is always related to the dimensions of the unit cell (the minimum repeat unit having the full symmetry of the whole pattern). The Bragg law expresses the fact that waves scattered from the atoms in successive planes reinforce one another to give an observable 'reflection' only when their path difference is an integral multiple of λ, the wavelength. The interplanar spacing d has a maximum value for any given atomic array, but strictly speaking, it has no minimum value. If one looks at an orchard or a plantation consisting of trees planted at regular intervals in all directions one sees rows of trees. There is a maximum distance between successive rows, but as the angle of vision becomes more and more oblique the rows seen appear to come closer and closer together indefinitely. In practice, however, there *is* a lower limit to the value of d that can lead to reinforcement of an incident wave by a reflected wave, and that lower limit is such that $2d = \lambda$; the incident beam is then at right angles to the set of scattering planes. The shorter the wavelength, the more detail we would expect to be able to observe, since we get more spots; but, as has already been pointed out, the intensity of scattering varies with λ and, in practice, only X-ray wavelengths between about 0·7 and 2·3 Å are really useful. The result of not being able to observe diffraction effects for values of d less than $\lambda/2$, especially when λ is fairly large (as, for example, when using the experimentally convenient CuK radiation), is that the deduced picture of the original structure obtained from this limited diffraction pattern tends to contain *false* detail. Mathematically, we would say that, whereas our structure analysis depends on the assumption that we are dealing with

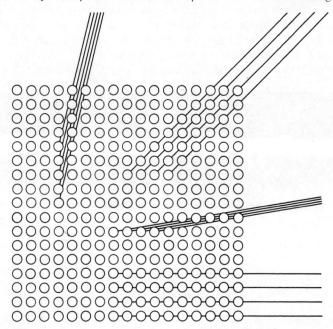

FIGURE 49 Spacing between parallel consecutive rows of atoms in different directions in a square network.

infinite series, in practice our series are finite, and we get false 'termination-of-series' effects.

The actual interpretation of the intensities of the diffraction effects observed is rendered even more difficult by what is often called 'the phase problem'. Measurement has given us what is effectively a reproduction of the periodic scattering power in different crystal orientations (Figure 50). What we do not know from the pattern alone is how to superimpose them. If we did, then crystal-structure analysis would become purely automatic and perhaps even a little dull. Theoretically, we could instruct a computer to superimpose them in every conceivable way and select the arrangement which is most plausible. If only a few measurements are involved this can indeed be done. (The Monte Carlo technique it is called, for obvious reasons!) Usually, however, we have hundreds or thousands of observations, and a Monte Carlo technique is not practicable. However, we are often able to make a few shrewd guesses, on the basis of our previous experience of atomic sizes, molecular groupings, etc.; and we can make use of obvious limitations, such as the fact that we *are* dealing with atoms, each of which occupies a finite space, and that the structure would

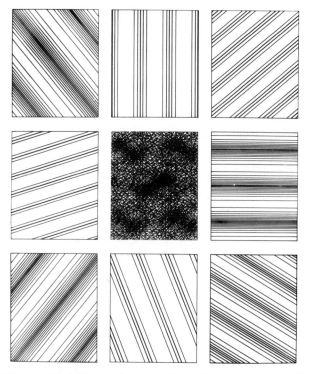

FIGURE 50 Diagram illustrating the 'phase problem'. X-ray measurements give, in any direction, the distribution of scattering power as a series of sine waves. But although amplitude and spacing are known, the phase (relative to some arbitrary origin) is not known. If the waves are correctly superimposed (as at the centre) the atomic positions emerge as peaks in the combined pattern. But a shift sideways of any wave would lose the peaks.

not hold together if the atoms were too far apart in any given direction. We can also use various mathematical tricks too sophisticated to be described in a book such as this, which depend on the fact that electron density cannot ever be negative. Another valuable device, in particular for the study of protein structures, is to make use of what is called *anomalous dispersion*. If the wavelength used is near to one of the resonance frequencies of some atom in the structure, then the waves are heavily absorbed (see p. 35) and the f factor is affected in such a way that the phase of the scattered wave can sometimes be deduced by considering the intensities of suitable pairs of reflections.

Our ultimate test of any structure we deduce is to calculate backwards and to see whether such a structure would give the diffraction pattern we observe: the trial-and-error technique. Once we have achieved an agreement which must be nearly right, we refine it by successive approximations. In all this calculation a computer is almost essential in order to minimise the time taken. However, crystal structure determination — seeing the way in which atoms (even the kinds of which may be unknown) are arranged in crystalline solids—is never likely to become purely automatic, since a computer does not give the right answers unless it is asked the right questions, and the questions to be asked depend on the problem to be solved.

Practical Considerations

Now so far we have said very little about the experimental techniques we use to obtain diffraction patterns from whatever kind of specimen is available. Von Laue's assistants were able to obtain a diffraction pattern using a stationary crystal because they had 'white' radiation (Figure 51). A

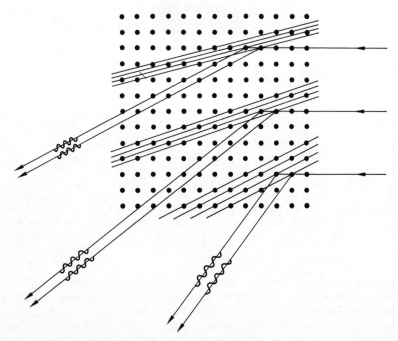

FIGURE 51 X-ray diffraction can take place from a stationary crystal provided that the incident beam contains a range of wavelengths. Reflection then takes place from the various atomic planes, as from a series of arbitrarily placed mirrors. (Contrast Figure 40, also by courtesy of Dr. Chas. Bunn.)

stationary crystal means that θ is fixed for all sets of planes and, since d is not adjustable, it must be λ that is variable for the Bragg law to be satisfied. Bragg's ionisation spectrometer and modern diffractometers use a single wavelength, but vary θ by turning the crystal (Figure 40). Modern photographic techniques are of several kinds, but except for very special purposes, they all use monochromatic radiation.

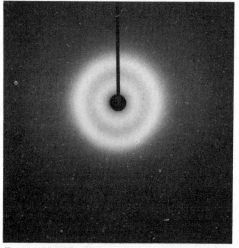

FIGURE 52 Diffraction pattern of polystyrene, an amorphous compound (Cu$K\alpha$ radiation).

If the specimen is an amorphous solid, then only a diffuse ring, or two or three diffuse rings, are obtained when diffraction of a fine beam of X-rays is recorded photographically (Figure 52). If the substance is in the form of a fine powder or a conglomerate of small crystallites (about 10^{-4} cm across each crystallite), then the diffraction consists of a set of sharp concentric cones. The semi-vertical angles of the cones θ_1, θ_2, etc., correspond to the crystal spacings d_1, d_2, etc. in accordance with the Bragg law $n\lambda = 2d \sin \theta$, λ being constant and n taking successive integral values. The actual pattern recorded depends upon the shape of the photographic film used (Figure 53). If the crystallites are very small indeed (10^{-5} cm or less), the 'powder' lines are broadened, and gradually for colloidal specimens of decreasing particle size the pattern begins to approach that of an amorphous solid. If the crystal grains are 10^{-3} cm across or larger the powder lines become spotty (Figure 54) and the specimen must be rotated for smooth rings to be obtained. Generally, diffraction patterns from powders are recorded in strip form (Figure 55), and these are used for purposes of identification. Three or four of the more intense lines of high spacing (large d), that is, of low angle θ, are usually enough for unambiguous identification to be achieved. This technique is graphically known as the finger-print method.

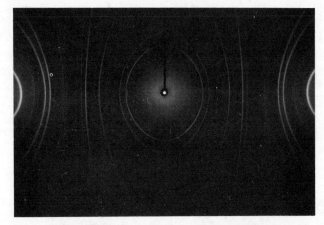

FIGURE 53 The section of a series of concentric cones by a cylinder placed with its axis normal to the common cone axes. Powder diagram of platinum, as registered on a cylindrical film afterwards straightened out (Cu$K\alpha$ radiation).

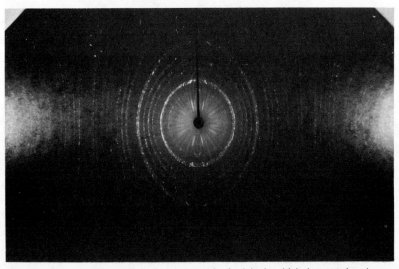

FIGURE 54 Spotty lines on a powder photograph of calcite in which the crystal grains are rather large (10^{-3} cm approximately). Rotation of the specimen would give smooth lines, by randomising crystallite orientations (Cu$K\alpha$ radiation).

FIGURE 55 Strip powder photograph of an experimental type of dental cement.

A single crystal allowed to rotate about a random direction which is not near to any prominent 'crystal axis' gives a pattern which approximates to that from a powder (Figure 56). Small single crystals are usually placed so that some prominent crystal axis is perpendicular to the collimated X-ray beam (usually 0·5 mm diameter) and the crystal is made to rotate, or to oscillate about a set of positions such that successive patterns overlap slightly. The photographic film is in a cylindrical holder with the axis of crystal rotation as the axis of the cylinder. If the film-holder is kept stationary the photograph obtained consists of straight (layer-) lines of

FIGURE 56 Diffraction pattern from a single crystal of hexa-
methylenetetramine which is rotating about an axis not
coincident with any major crystal axis. (Random rotation photograph.) The spots obviously lie on 'powder
rings' (CuKα radiation).

diffraction spots when the film is straightened out (Figure 57). Measurement of spot position and photometered intensity is simple, but for complicated structures with large unit cells, this technique gives insufficient resolution and it is usually replaced by a moving-film method. Two of these are very well known. In the Weissenberg method there are cylindrical screens inside the film-holder which allow only one layer-line to pass through them at a time. The screens are stationary, but the film-holder moves to and fro behind them, parallel to the rotation axis and synchronously with the rotation of the crystal. Distances along the film can then be correlated with angles of rotation, and the diffraction spots on a single layer-line are spread out over the whole film (Figure 58). The resolution is good and interpretation is not difficult, but as the diffraction spots are arranged in festoons, intensity measurement with a photometer is not quite so easy as when they are along straight lines. The second moving-film technique, the precession method, which was introduced by M. Buerger of

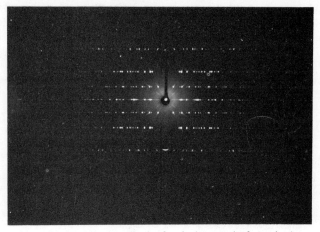

FIGURE 57 Rotation photograph of a single crystal of struvite (magnesium ammonium phosphate) dissected out of a bladder stone. Struvite crystals are also found in the bones of decaying tinned fish (Cu$K\alpha$ radiation).

FIGURE 58 Weissenberg photograph of a single crystal of hexamethonium bromide, a compound which produces paralysis in the nerve system (Cu$K\alpha$ radiation).

M.I.T., involves a plane film and a much more complicated film movement behind a screen. It does give straight lines of spots (Figure 59)—in fact, the patterns obtained are the nearest in appearance to Lipson's optical diffraction patterns—but the use of a plane film-holder means that the cut-off is not at $2\theta = 180°$ but more nearly at $2\theta = 60°$. The method is widely used for substances such as proteins and viruses with very large unit cells.

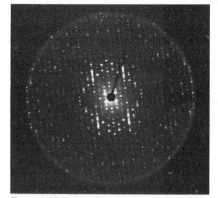

FIGURE 59 Precession photograph of a haemoglobin crystal. (Compare Figure 44.) (By courtesy of Dr. Max Perutz.)

Finally, there are the fibrous and platy substances. The fibres in a filament of silk or of feather have one axis in common, but otherwise the structures are random. Similarly, the arrangements of atoms in some kinds of commercial carbon are like those in graphite but much more limited in extent and with only their normals in common. Provided that there is one axis in common, however, the dif-

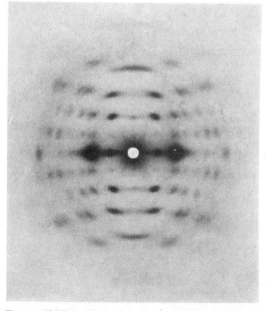

FIGURE 60 Fibre photograph (due to the late Professor W. T. Astbury, courtesy of *Endeavour*) of artificial silk, which is relatively well organised.

fraction pattern observed photographically when that direction is normal to the X-ray beam and parallel to the photographic film is very similar to that from a single rotating crystal (Figure 60), although usually the spots are spread out because of the smallness of the individual micelles and because of some lack of parallelism.

To sum up: X-rays are diffracted by atomic groupings because the wavelengths of X-rays and the atomic separations are comparable. When the atoms form a crystalline solid the regularity of repetition of the scattering centres makes the crystal an ideal three-dimensional diffraction grating for waves of about $0 \cdot 7 - 2 \cdot 3$ Å in length. This range covers the X-rays characteristic of targets made of molybdenum, copper, nickel, cobalt, iron and chromium. Care must be taken that these soft rays are not allowed to come into contact with eyes or skin, as this would cause a dangerous burn through absorption. Various experimental techniques are used for obtaining the diffraction spectra from different types of specimen, which may be a fine powder, a conglomerate such as a metal wire or sheet, a single crystal usually less than $0 \cdot 5$ mm linear size or a fibre. Diffraction spots are obtained from single crystals, lines from powders, and diffuse patches, streaks and rings correspond to the imperfections introduced by the inevitable thermal vibration and by various kinds of structural disorder culminating in the limiting disorder of amorphous solids, liquids and gases. Interpretation is seldom, if ever, automatic, but is almost always assisted by automation. X-ray crystallographers are, in fact, among the major scientific users of computers.

SIX

MICROWAVES

1. Introduction

The term 'microwave' came into general use among scientists during the last war. It was used, and is still used, to denote an electromagnetic wave of a few centimetres in wavelength. Although the term is relatively modern, experiments with microwaves were carried out before the year 1900. The resurgence of interest in microwaves during the 1940s arose from the great potential value for radar of this region of the electromagnetic spectrum, and it is quite natural that a convenient name for these waves should have found wide currency at that time.

It is perhaps surprising to find the prefix 'micro' describing waves which are not, after all, so very short by terrestrial standards. The explanation lies in the contrast with the very much longer electromagnetic waves previously used in radar and communications systems. Wavelengths measured in metres, tens of metres and hundreds of metres were commonly used; in comparison, centimetric radio waves seemed very short indeed, and the term 'microwaves' was accepted as entirely appropriate. However, it is important to keep a sense of perspective in discussing a particular part of the electromagnetic spectrum, and it may help to bear in mind that microwaves are some 100,000 times *longer* than light waves!

Just as light waves at the red end of the spectrum merge imperceptibly into the infra-red region, so short radio waves merge into the microwave region with no natural boundary between the two. For practical purposes, such as international agreements or nomenclature, a precise wavelength boundary may be set, but the value of this is quite arbitrary. Its choice is purely a matter of convenience, and has no fundamental significance.

For many years there existed an unexplored range of wavelengths between infra-red waves and microwaves. It was unexplored simply because suitable sources and detectors did not exist. This wavelength gap became narrower and narrower as microwave and infra-red techniques improved. The quantum electronics revolution has recently produced the infra-red laser, and microwave electronics has produced the backward-wave oscillator and the crystal harmonic generator. These sources now overlap in wavelength, and in this sense the gap between infra-red waves and microwaves no longer exists. But it is important to note that continuous coverage of this region of the electromagnetic spectrum has not been achieved at the time of writing. Infra-red lasers are tunable over a very small wavelength range only, in contrast to the backward-wave oscillator in which a substantial variation of wavelength can be obtained by simply altering the voltage applied to it. The situation at present is that, at wavelengths from about one-third of a millimetre upwards, any desired wavelength can be generated by means of a range of backward-wave oscillators. (Not all of these are necessarily available commercially.) From a third of a millimetre and extending down to the optical range of wavelengths, a large number of discrete wavelengths can be generated by various infra-red and optical lasers.

In what follows we shall describe the development of microwave techniques, following a roughly chronological order. We shall pay special attention to the physical principles underlying the operation of the klystron and the backward-wave oscillator. These are perhaps the most important microwave sources, especially at very short wavelengths. We shall then discuss an application of microwaves to large-scale measurement of distance, showing how an airborne microwave radar can be used to produce a map of good quality automatically, in no more time than the aircraft takes to fly over the region to be mapped.

2. Early Researches

The most important single step in electromagnetic theory was, of course, Maxwell's statement of the fundamental equations and his deduction from these that light is electromagnetic in character. An account of the historical significance of Maxwell's work was mentioned in Chapter One, and it is enough for our purpose now to note the following features of Maxwell's theory:

  (i) It identified light with electromagnetic waves of extremely short wavelength.
 (ii) It gave the correct numerical value for the velocity of light.
(iii) It predicted the existence of electromagnetic waves of arbitrarily long wavelength.

Maxwell published his theory in 1865, little more than thirty years after Faraday's discovery of electromagnetic induction. Its publication encouraged experimentalists to look for these longer electromagnetic waves. Hertz was successful in generating and detecting electromagnetic waves by electrical means (see p. 14). In 1888 he measured the velocity of his electromagnetic waves and, although his determination was inaccurate, he showed that the velocity was of the same order as the velocity of light. Hertz also showed that his electromagnetic waves travelled more or less in straight-line paths, allowing for diffraction. For his wavelength of 66 cm diffraction effects would be very important, as we shall see later. A characteristic property of light waves is the phenomenon of polarisation: Hertz designed ingenious experiments to show that electromagnetic waves were also polarised. He used parabolic-cylindrical mirrors to focus his waves, and a prism of pitch to refract them. Having established so many common features between light and electromagnetic waves, he wrote,

'It is a fascinating idea that the processes in air which we have been investigating represent to us on a millionfold larger scale the same processes which go on in the neighbourhood of a Fresnel mirror or between the glass plates used for exhibiting Newton's rings.'

Hertz's experiments were repeated and extended by many subsequent workers, among them Righi, Trouton, Lodge, Fleming and, of course, Marconi. But especially interesting for us is the work of J. C. Bose of Calcutta. In 1897 Bose used a cylindrical reflector to form beams of radiation at a wavelength of about 5 mm. With this very short wavelength,

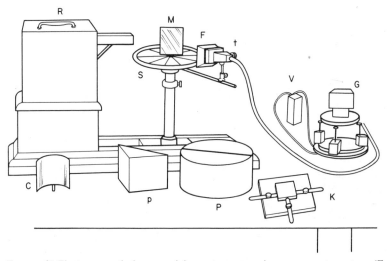

FIGURE 61 Electromagnetic horn receiving antenna on microwave spectrometers. (From Ramsay in Proc. I.R.E., Feb. 1958, Vol. 46, No. 2 and by courtesy of Longman's Green & Co.)

diffraction effects would not be too serious, and the similarity of his waves and light waves would be greater for this reason. Bose followed up the optical analogy with great skill, and the microwave spectrometer he used for quasi-optical demonstrations is shown in Figure 61. This is a striking example of the remarkable progress he made in spite of the very inadequate sources and detectors available at that time.

Work in this field would certainly have proceeded more rapidly had better sources been available. All the early workers used spark-gap oscillators of various kinds. Bose used a platinum sphere as his resonant element, with two smaller metallic spheres, diametrically opposite and close to the large sphere. The small spheres were connected to an induction coil, and, when this was energised, sparks occurred between the small spheres and the large sphere. Thus a current could be passed suddenly through the large sphere. The transient response of the sphere would die out quite rapidly, but in an oscillatory manner with a fairly well-defined frequency. The sphere would therefore behave like an oscillating electric dipole and electromagnetic waves would be radiated. However, such a source was not satisfactory for accurate work. It was erratic and its frequency spectrum was broad.

The next important step was the invention of the thermionic valve by J. A. Fleming in 1902, coupled with the insertion of a grid by Lee de Forest. The resulting triode valve formed an excellent amplifier and could be used as an oscillator by making use of a suitable tuned circuit. However, the triode valve was not suitable for the generation of very short waves. In any case, technical interest was moving towards longer wavelengths, as these seemed to be more useful for radio communication. For these reasons there was a period of relative inactivity in microwaves until 1936. Nevertheless, some important work was done during this period— Marconi's pioneering work on line-of-sight communication links and on tropospheric scatter propagation as early as 1930, and the Clavier–Gallant 17-cm wavelength cross-channel link are examples.

In the 1930s many workers gave thought to the problem of using electron tubes to generate very high frequencies. There were several difficulties which prevented the use of the conventional triode valve at very high frequencies. A fundamental difficulty was the finite time taken for an electron to pass from cathode to grid and from grid to anode. The simple theory of the triode ignores this transit time. This is a valid assumption provided that the transit time is much less than the period of the oscillation to be generated. An obvious way to reduce transit time is to reduce the cathode/grid and grid/anode spacings, but equally obviously there are practical difficulties in making triodes with extremely small spacings. Nevertheless, in the 1930s triode valves were commercially available which could be used in oscillators at frequencies as high as 300 MHz, corresponding to a wavelength of 100 cm. Moreover, new ideas were being aired, and it was clear that, although the conventional triode valve had inherent limitations as a microwave source, it by no means followed that *all* electronic devices would suffer from the same limitations. It was reasonable to assume that further research would produce electronic oscillators for wavelengths much shorter than 100 cm. If such sources became available other problems would arise, notably that of connecting the oscillator to the aerial in a radio-communication system. It was expected that conventional two-wire transmission lines and concentric cables would not be suitable because of their high attenuation (waste of power in the form of heat) at very high frequencies. Two groups of workers had the foresight to tackle this problem in anticipation of the eventual availability of suitable sources; one group at M.I.T. was led by Barrow and Chu, the other at Bell Telephone Laboratories led by Southworth and Schelkunoff. The starting-point for this work was a paper published in 1897 by Lord Rayleigh in which he showed that electromagnetic waves could, under certain conditions, be propagated through hollow metal pipes. Although Lodge and Fleming, in their early experiments with spark-gap microwave sources, had made some use of hollow metal pipes, they had not carried out a systematic experimental verification of Rayleigh's theoretical work. Indeed, it would have been quite impossible to verify the theory accurately with such sources; it has already been said that the spectrum of frequencies generated was rather broad, and precise verification of the theory would need a much better approximation to the pure monochromatic source assumed in the theory.

The advent of the electronic triode valve led to the development of valve oscillators of excellent spectral characteristics and, for the purpose of verifying Rayleigh's theory, the triode valve was certainly entirely satisfactory. The only disadvantage was that the shortest available wavelength was still rather long for convenience, as Rayleigh's theory predicted that propagation of waves inside a pipe could occur only if the diameter of the pipe were comparable with the wavelength. The researches of Barrow, Chu, Southworth and Schelkunoff verified all the conclusions reached by Rayleigh in a most convincing manner, and the feasibility of hollow-pipe transmission was firmly established. However, Rayleigh's papers considered perfectly conducting tubes only, and a practical question was how serious the effect of finite conductivity would be. Theoretical researches at M.I.T. and at Bell settled this question in a satisfactory manner; the attenuation of hollow-pipe transmission systems, or 'waveguides' as they are now known, proved to be satisfactorily small—considerably smaller than the attenuation in a concentric cable. This conclusion was soon verified experimentally, and the waveguide was firmly established as the microwave equivalent of the two-wire line or concentric cable.

It is important to notice that, although the first approach to microwaves by Hertz, Bose and others was made from the optical standpoint, using optical analogies to guide the experimental work, Southworth, Barrow and their collaborators had quite a different approach. In the intervening period of time communications engineering had developed rapidly, and there was in existence a substantial body of theoretical analysis and experimental verification of the behaviour of high-frequency transmission lines and aerials. It was understandable that those who foresaw the potential value of microwaves to the communications engineer also wished to exploit the existing body of knowledge of high-frequency systems in developing analogous techniques for microwaves.

We must now look at the two approaches in more detail. When we do this we find a fundamental physical law which tells us the limits of validity of the optical viewpoint.

3. Aerials, Waveguides and Cavity Resonators

We have already seen that electromagnetic waves can be guided by hollow metal pipes and that waveguides of this kind can be used to transmit microwave energy from one point to another, just as conventional transmission lines, such as the concentric cable, are used at longer wavelengths.

When we approach the problem of transmitting electromagnetic energy efficiently from the telephone engineer's standpoint it is quite natural to take for granted that a physical connection must be made between the sending equipment and the receiving equipment. On the other hand, looking at the matter optically, such a connection does not seem necessary. It is a simple matter to form a parallel beam of light by means of a lens or a parabolic mirror, and if this is done, all the light energy sent can be

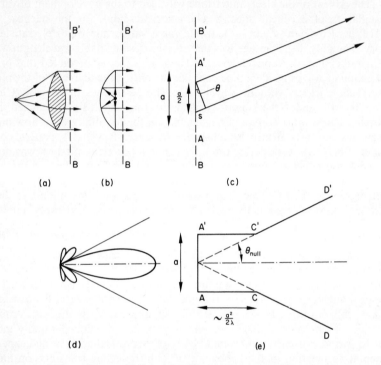

FIGURE 62 Formation of a beam of radiation.

received by a collector of the appropriate size. A physical connection is not necessary in this case; why is it necessary for the longer electromagnetic waves we call microwaves? To answer this question we need to carry out a very simple but very important theoretical calculation. Figure 62 (a) shows how a lens can be used to form a parallel beam of light from a point light source, and Figure 62 (b) shows how a paraboloidal mirror can be used for the same purpose. Both schemes have the characteristic that the optical path length from the source to a point on the plane BB' is the same for all possible paths. Thus BB' is a wave front, or equiphase surface. In the case of the paraboloid all ray paths from the source to the plane BB' have the same *geometrical* length; in the case of the lens the velocity of propagation is slower in the lens itself, and this effectively lengthens the geometrically shorter paths to be *optically* equal to the geometrically longer paths.

So far, this argument would seem to predict parallel-straight-line propagation to the right of *BB'* for an indefinite distance. This is not the case, however. Ray theory, which works fairly well at small distances, fails completely at great distances. We shall now try to explain the nature of this failure. To do this we take a two-dimensional model for simplicity. Referring to Figure 62 (c), we assume that the plane *BB'* is illuminated by a wave of uniform phase and amplitude over the region *AA'*. According to Huygens' principle (discussed earlier in Chapter 1), the field at a given point to the right of *BB'* can be calculated by considering the region *AA'* split into narrow strips and adding the contributions from the individual strips with the appropriate time delay to allow for the different path lengths from each in the chosen direction of propagation.

We shall not carry out the full calculation, which is done in most books on microwave aerials or optics, but we shall show that there is no radiation at all at certain definite angles (see also Chapter 1, p. 18). To do this, we need only divide the region *AA'* into two equal parts. The distance from the centres of these two parts to a very distant point in the direction θ will differ by the distance *S*, as shown in Figure 62 (c). By simple trigonometry,

$$S = \frac{a}{2} \sin \theta \qquad (1)$$

The condition for no radiation requires path differences of $\frac{\lambda}{2}$, $\frac{3\lambda}{2}$, $\frac{5\lambda}{2}$, etc. In this case a wave crest from one half of *AA'* coincides with a trough from the other half.

This leads to

$$\sin \theta_{\text{null}} = \pm \frac{(2n + 1)\lambda}{a} \qquad (2)$$

or

$$\sin \theta_{\text{null}} = \pm \frac{\lambda}{a}, \pm \frac{3\lambda}{a}, \pm \frac{5\lambda}{a}, \dots \text{etc.} \qquad (3)$$

In general, therefore, several nulls will be found in the radiation pattern to the right of *BB'*, corresponding to the several possible choices of *n*.

All integral values of *n* are possible so long as $\sin \theta_{\text{null}}$ is not greater than unity. Let us calculate the *smallest* angles at which the radiation falls to zero for the special case of an aperture 10 wavelengths across. From (3) we find

$$\sin \theta_{\text{null}} = \pm 0\cdot 1$$

or

$$\theta_{\text{null}} = \pm 5\cdot 7$$

for

$$a = 10\lambda$$

similarly,

$$\theta_{\text{null}} = \pm 0\cdot 57^{\circ}$$

for

$$a = 100\lambda$$

The first calculation above would apply, for example, to an aperture of 10 cm width at a wavelength of 1 cm, and the second calculation to an aperture of 100 cm at the same wavelength. Thus an aerial of quite manageable size can produce a narrow beam of radiation when operated at the short wavelengths we call microwaves.

There are several practical forms of narrow-beam microwave aerial based on this principle—the lens and the paraboloid are two possibilities. Yet another is the electromagnetic horn; a metal cone or pyramid fed with electromagnetic energy at its apex and radiating from its open base. Such aerials are to be seen on high buildings and towers in most highly developed countries, where they play an important role in television relay circuits. They are also important in radar, though the parabolic cylinder is also frequently encountered and has the advantage that the vertical and horizontal apertures can be chosen separately; thus, the vertical and horizontal beam widths can be controlled independently.

The possibility of producing very narrow beams of radiation from aerials of reasonable size is essential to radar, and it was the urgent need to develop radar with high angular discrimination for military purposes that led to the rapid development of microwave techniques during the war years.

The general form of the radiation pattern is sketched in Figure 62 (d), and it is clear that it is quite different in form from the uniform parallel-sided beam which is found near the source.

We next consider the transition from the near-field region to the far-field region. The simplified diagram of Figure 62 (e) illustrates the main features. In the near-field region between AA' and CC' we have a parallel-sided beam in which the intensity falls rather sharply to zero on AC and $A'C'$. Beyond CC we enter the far-field region in which the beam occupies a finite width in *angle*. $C'D'$ and CD are drawn to coincide with the first zeros of the far-field pattern, so that the main beam lies within these limits.

The lines ACD and $A'C'D'$ now delineate roughly the transition from the near-field to the far-field region. We can calculate quite easily the position of the plane CC' which divides the near-field from the far-field region. Putting $n = 0$ in (3) we have for the angle θ_{null} the equation

$$\sin \theta_{null} = \frac{\lambda}{a} \qquad (4)$$

We also have, by elementary trigonometry,

$$\tan \theta_{null} = \frac{\frac{1}{2}a}{AC} \qquad (5)$$

If $\lambda \ll a$ so that $\sin \theta_{null}$ is small, we have $\tan \theta_{null} = \sin \theta_{null}$, and it then follows that

$$AC = \frac{a^2}{2\lambda} = R, \text{ say} \qquad (6)$$

This distance R is often called the 'Rayleigh' range, and separates the near-field region from the Fraunhofer or far-field region.

We are now in a position to answer the question raised at the beginning of this section; namely, why is it necessary to provide some means of guiding microwaves from one point to another for efficient transmission of energy, whereas with light no such guide is necessary?

Let us suppose that we have an initial beam width of 10 cm; this corresponds to AA' in Figure 62 (e). Let us also suppose a wavelength of 1 cm. Substituting in equation (6), we find

$$R_{microwaves} = 50 \text{ cm}$$

For light waves we can take a wavelength of 5,000 Å, or 5×10^{-5} cm. With the same initial beam width we now find

$$R_{light} = 10^6 \text{ cm} = 10 \text{ km}$$

Thus, for microwaves, diffraction leads to significant beam spreading on the *laboratory* scale of distances, whereas for light beam spreading is significant only on the *terrestrial* scale. Thus, a substantially parallel beam of light can easily be maintained from one end of a laboratory to the other with an aperture of only 10 cm, and can be collected by an aperture of the same size with virtually 100 per cent efficiency. On the other hand, a microwave beam of 1 cm wavelength would suffer a great reduction in intensity due to angular spreading or diffraction of the wave energy.

An obvious way to stop this sideways spreading is to confine the waves in a hollow pipe. We can visualise this guiding effect by thinking of a ray reflected alternately at by two parallel mirrors as in Figure 63 (a). From an optical viewpoint, it would appear that the ray

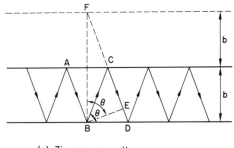

(a) Zig-zag ray path

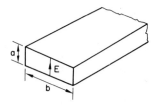

(b) Rectangular waveguide

FIGURE 63 Waveguide theory.

could be arranged to make any desired angle θ with the mirrors, and this would be the case for a very narrow *light* beam or 'ray' in which diffraction effects could be neglected. As the wavelength increases towards the microwave range of wavelengths, diffraction effects become important, and energy will begin to spread out sideways from the ray as it follows its zigzag path between the mirrors. Eventually energy will spread out from one ray to the adjacent parallel ray and interference effects will take place. Provided that the waves interfere *constructively*, propagation can still take place, but it turns out that this is possible only at a finite number of possible ray angles. In the special case in which the mirror separation is greater than a half-wavelength but less than a whole wavelength only *one* ray angle is permissible. We then say that only one *mode* of propagation is possible, and this mode is called the *dominant* mode.

These statements are quite easily proved with reference to Figure 63 (a), in which certain construction lines have been added; these are shown as broken lines in the diagram. If the waves represented by the rays AB and CD are to interfere constructively when lateral spreading takes place BE must be a wavefront common to the two rays. This can be the case only if the distance $BC + CE$ is a *whole* number of wavelengths, $n\lambda$ say. It is clear from the diagram that $BC = FC$, however, so that we must have $FE = n\lambda$. But $FE = \sin \theta$, and noting that $FB = 2b$, we see that

$$\sin \theta = \frac{n\lambda}{2b} \qquad (7)$$

Provided that $\frac{\lambda}{2b} < 1$, there will be at least one solution of (7); if $\frac{\lambda}{2b} \ll 1$ there will be many. Thus, we have proved the earlier statement that only a finite number of ray angles is permitted.

Now let us consider the case in which the mirror separation lies between $\frac{1}{2}\lambda$ and λ. Then $\sin \theta$ lies between n *and* $\frac{1}{2}n$. As $\sin \theta$ must be less than unity, we must have $n = 1$, and there is *only one possible value of θ*.

So far we have considered a beam spreading in one dimension only; it is obvious that, by using *four* reflecting surfaces in the form of a hollow rectangular tube, as shown in Figure 63 (b), the wave is completely trapped within the tube and cannot spread in any direction. The possible ray paths are now much more complicated, however. A ray can be reflected in turn at each of the four walls, corkscrewing its way down the tube. In this case *two* integers, m and n, are needed to define the modes of propagation. One of these deals with the phase condition for the side a, and the other deals with the phase condition for the side b, as before.

We implicitly assumed earlier that the integer n could not have the value zero. It is, however, possible for *one* of the two integers m and n to have this value. For example, in the case shown in Figure 63 (b), with the electric field following straight lines between the top and bottom faces of the tube, we can have $m = 0$, $n = 1$.

If a is less than $\lambda/2$ and b lies between $\lambda/2$ and λ this is the only possible mode, and is called the *dominant* mode. Because the magnetic field has a component parallel to the axis of the tube, the mode is called the H_{01} mode and is the most important in microwave engineering. (When waves are trapped inside a metal tube the directions of the electric and magnetic fields are more complex than those shown in Figure 8.) Hollow metal tubes used in this way are called *waveguides*, and there is a mass of literature on the theory and application of these guides. We can think of a tubular waveguide for electromagnetic waves as analogous to a speaking tube for sound waves. In both cases energy is transmitted quite effectively from one end of the tube to the other because the tube prevents the sideways spread of energy which would otherwise take place.

There is another useful analogy between sound waves and microwaves. Just as an organ pipe gives a strong resonance for a certain note or wavelength of sound, so an electromagnetic resonator can be devised to respond to a particular wavelength. A waveguide, closed at both ends to prevent the escape of energy, is an example; of course, a small hole must be provided to couple the resonator so formed to the waves with which it is to

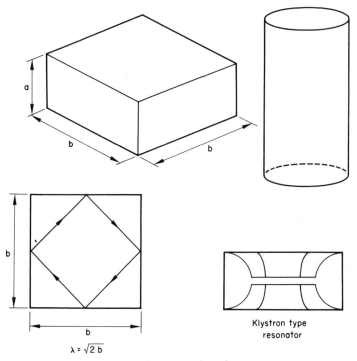

$$\lambda = \sqrt{2\,b}$$

Klystron type
resonator

FIGURE 64 Electromagnetic cavity resonators.

resonate. Sound resonators can take many forms: the organ pipe is one, but an empty bottle and a policeman's whistle are two alternative forms. Electromagnetic resonators also have many forms. Several of these are shown in Figure 64, and they are called *cavity resonators*. It is possible to calculate the wavelength at which a given resonator will respond, the calculation being quite easy for simple geometrical shapes, but very difficult in many cases. A particularly simple case is the rectangular resonator shown in Figure 64. This can be regarded either as a rectangular waveguide parallel to the X axis, closed at each end by walls parallel to the Y axis, or as a waveguide parallel to the Y axis, closed at each end by walls parallel to the X axis. As the cavity is *square* in plan, a possible mode has $\theta = 45°$, and we can find the wavelength for resonance by putting this value into the waveguide formula (7). The *longest* wavelength for resonance is found by putting $n = 1$, and works out to be

$$\lambda = \sqrt{2}b \qquad (8)$$

Although this simple equation is valid only for the particular case of the longest wavelength mode of a rectangular cavity, it is true in general for *any* mode in *any* cavity that the wavelength is directly proportional to the linear dimensions of the cavity.

Thus, for example, if we wish to *halve* the wavelength for resonance we must *halve* all the linear dimensions of the cavity.

This is an important point because, if we wish to use very short wavelengths, the resonator dimensions will be very small. We shall see in the next section that this is one of the factors which makes the generation of millimetric electromagnetic waves a difficult problem.

4. Generation of Microwaves

We have already drawn an analogy between an organ pipe and an electromagnetic cavity resonator. The analogy can be pushed a little further. When a stream of air is blown through the organ pipe its natural resonance is excited and maintained; a continuous oscillatory sound wave is generated. When a stream of electrons passes through a suitably designed cavity resonator, electromagnetic oscillations may be generated continuously. However, the analogy between sound waves and electromagnetic waves is not close enough to allow a more detailed comparison, and at this point we must adopt a more fundamental point of view in order to understand the operation of a microwave oscillator. A simple oscillator consists essentially of a resonator, a source of power and a synchronising system which controls the source of power and supplies power to the resonator at the correct point in the cycle.

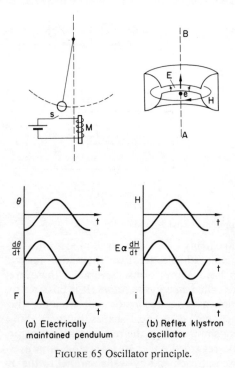

(a) Electrically
maintained pendulum

(b) Reflex klystron
oscillator

FIGURE 65 Oscillator principle.

For example, Figure 65 (a) shows an electromagnetically maintained pendulum. This is essentially an oscillator in which the resonator is the pendulum, the source of power is the battery and the synchronising system (not shown) might be a photo-electric device operated by the pendulum and controlling the opening and closing of the switch S. The pendulum defines the frequency of oscillation. As the pendulum approaches the electro-magnet M the switch is momentarily closed, and an attractive force is applied for a short time. This is arranged to occur during each *forward* swing of the pendulum, and in this way pulses of power are supplied to the pendulum in such a way as to make up for frictional losses and keep the pendulum swinging. Note that the sustaining impulse is best applied when the angular velocity has its maximum value.

Now consider the problem of maintaining oscillations in an electro-magnetic cavity resonator (Figure 65 (b)). The mode of oscillation of the resonator with which we are concerned is circularly symmetrical, the magnetic field lines being circles, as shown. The electric field is strong in the central region only, and is proportional to the rate of change of the magnetic field—this was discovered by Faraday and is known as Faraday's law. If a tightly bunched group of electrons is shot through the cavity at the time when the electric field is a maximum, and in such a direction as to slow the electrons down, the reduction in kinetic energy of the electrons will be delivered to the electromagnetic field of the cavity. If this process can be repeated synchronously at the correct point in each cycle oscillations can be maintained.

The production of a suitably bunched electron beam from an initially uniform beam is achieved very neatly in a microwave oscillator known as the reflex klystron. An initially uniform unbunched beam of electrons is shot through the cavity from A to B (see Figure 65 (b)). A negative electrode at B repels the electrons, driving them back through the hole in the cavity resonator again. An electron which passes through the resonator from A to B when the field is in the accelerating direction *gains energy* from the field, but an electron passing through half a cycle later *loses energy* to the field. There is no *net* transfer of energy to the resonator from the electrons which are moving from A to B because this stream of electrons is uniform and as many electrons gain energy as lose energy. However, the faster electrons travel deeper into the retarding field set up by the negative electrode at B, and so take a longer time to return to the resonator than those which make their first passage through the resonator half a cycle later. If the retarding potential is correctly chosen the fast and slow groups of electrons return to the resonator in a bunch at about the same time. If at this time the electric field acts so as to *retard* the electrons energy is transferred to the electromagnetic field and the resonator gains energy. Because of the bunching action, the loss of energy from the field to the electrons half a cycle later is very much less than the energy gained by the field when a bunch passes through, and the required *net* transfer of oscillatory energy is obtained.

The bunching process can perhaps be visualised more clearly with the help of another analogy. Suppose two balls are thrown vertically into the air, one after the other. If the first ball is thrown harder it will go higher and will take longer to return to earth than the second ball. It is obviously possible to arrange matters so that both balls return to earth at the same time. Extending the argument, if balls are thrown at one-second intervals, alternately fast and slow, they will, if the strength of the throws is properly adjusted, fall to earth in *pairs* at *two-second* intervals. We can go further, and suppose that half a second after throwing a fast or slow ball we throw a ball at an average speed, so that at *half-second* intervals the sequence is fast, average, slow, average, etc. The balls return to earth in groups of four at two-second intervals. In the limit, a steady stream of vertically thrown balls can be arranged to return to earth in groups. These groups of balls could be used to maintain oscillations in a mechanical resonator such as a tray on a spring if its natural period of oscillation is two seconds. The analogy with the reflex klystron is now quite close. In the reflex klystron it is the *returning* electrons which maintain oscillations in the electro-magnetic cavity resonator.

Now let us consider what has been achieved in practice using the reflex klystron principle. The first reflex klystrons were made during the war years, and operated at a wavelength of the order of 10 cm. With an accelerating voltage of about 1,500 V for the electron stream, a power output of several hundred milliwatts was readily obtained. Equation (8) suggests that a rectangular cavity resonator 7 cm square could be used; the actual cavity used was similar in shape to that shown in Figure 65 (b), and about 5 cm in diameter.

Reflex klystrons for 3 cm wavelength soon followed, and then for 1·25 cm. Klystrons are now available for even shorter wavelengths, e.g. 8 mm, 4 mm, 2·5 mm. The cavity resonators for these klystrons are, of course, successively smaller—the resonator for a 2·5-mm klystron is minute, and the manufacture of such small resonators is a major difficulty in proceeding to even shorter wavelengths.

Even if this problem can be satisfactorily solved, however, another difficulty remains. This concerns the question of the formation of an electron beam of very high current density, and it arises in the following way. The size of the holes in the cavity resonator must be kept reasonably small in relation to the wavelength if loss of electromagnetic energy by unwanted radiation through these holes is to be prevented. To maintain constant performance in this respect, the hole diameter must be scaled down in proportion to the wavelength. The *area* through which the current must pass is therefore proportional to the *square* of the wavelength. If the total current required were constant a decrease in wavelength by a factor of ten would mean an increase in current density by a factor of a hundred. The situation is actually worse than this when other factors are taken into account. The difficulty of obtaining a very high current density is that the electrons tend to repel one another, and an electron beam tends to spread as it goes along because of this mutual repulsion effect. A high accelerating voltage is needed, and for various reasons this tends to be larger at shorter wavelengths. The power in the electron beam may be many tens of watts, and as the output is only a few tens of milliwatts, most of the electron beam power is dissipated as heat, developed when the electrons finally strike the cavity walls. There is therefore a big problem in getting the heat away from the cavity, and this is aggravated at short wavelengths by the very small size of the cavity structure. It is difficult to predict ultimate limits, but it seems unlikely that wavelengths much less than 1 mm will be generated by means of the reflex klystron.

In recent years an entirely different type of oscillator has been used most successfully for the generation of wavelengths as short as 0·4 mm. Karp in the U.S.A. and Convert in France have been in the forefront of

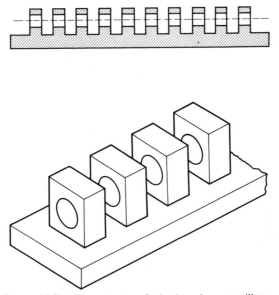

FIGURE 66 Slow-wave structure for backward-wave oscillator.

these developments, which are based on the development of a special device called a backward-wave oscillator. In this device, as developed by Convert, an electron beam is made to pass through a series of holes in a comb-like structure as shown in Figure 66. Interaction can take place between this electron stream and an electromagnetic wave travelling along the structure, and oscillations can be maintained. The comb acts like a waveguide for electromagnetic waves, although it is an open structure. It is only possible for electromagnetic waves to be guided in this way if the wave pattern is slowed down by the structure to a velocity considerably less than the velocity of light—the comb-like structure causes the wave to 'drag its feet', and the velocity may easily be reduced to one-tenth of the velocity of light in this way. The field structure of a slow wave is essentially that of a surface wave, and here another analogy may help. Waves on the sea do not greatly affect the water some distance below the surface simply because they are slow waves—slow, that is to say, in relation to the velocity of sound in water. At a depth below the surface of one wavelength the wave amplitude is very small indeed; the decrease in amplitude with depth follows an exponential law. It is for this reason that a submarine submerges in a storm. In the same way the field of the electromagnetic wave guided by a comb structure decays roughly exponentially with distance from the structure, and is very small indeed at a distance of one wavelength. The wavelength referred to here is that corresponding to the reduced velocity of the wave, and so may be only a tenth of the free-space wavelength. Thus, at millimetric wavelengths the field is extremely closely confined to the immediate vicinity of the comb, and the guiding is very effective.

It is quite a simple matter to accelerate an electron beam to a velocity of the order of one-tenth of the velocity of light, so that it is possible to keep the electrons exactly in step with the wave, and a cumulative interaction is possible. However, the comb structure has a peculiar feature, quite unlike the more usual kinds of waveguide we have so far considered, in that power flows along the structure in the opposite direction to the movement of the wave pattern. There is therefore an inherent feedback of energy from output to input, and under these conditions oscillations can be maintained. However, the wave-pattern velocity has a pronounced frequency dependence, and oscillations at a given frequency will occur only if the electron beam velocity is equal to the wave-pattern velocity at that frequency.

This proves to be a valuable effect because the backward-wave oscillator will oscillate at the frequency appropriate to the velocity of the electrons, and this is easily controlled by altering the voltage used to accelerate the electrons.

As there is no resonant cavity, the range of frequencies which can be generated by a single backward-wave oscillator is very great, a 2 : 1 range of frequency being feasible at the lower microwave frequencies. There is, however, quite a large variation in power output over the operating range of the oscillator. At millimetric and sub-millimetric wavelengths the tuning range is less, but 10 per cent range has been achieved.

Technological improvements have led to shorter and shorter wavelengths being generated by the backward-wave oscillator. It is interesting and important to know if there is any fundamental limit to this process. There is, first, the difficulty of making the comb structure. In a tube designed for 2 mm wavelength the teeth of the comb are only 0·12 mm thick and the gaps between the teeth are 0·1 mm. The hole through which the electron beam passes is only 0·4 mm in diameter. Moreover, these dimensions must be maintained all along the structure to a high degree of accuracy. For a shorter wavelength proportionately smaller dimensions will be required.

The other major difficulty is to obtain a high enough current on the electron beam. If the wavelength is *halved* the diameter of the hole through which the beam passes is also halved, and only a *quarter* of the area is available. Moreover, as the wavelength decreases, the length of an electron bunch must also decrease in proportion. For a given current density, therefore, the number of electrons per bunch in the beam will be roughly one-*eighth* of the number in a tube designed for twice the wavelength. For example, in one of Karp's tubes oscillating at just over 1 mm wavelength he estimated that there would be about 1,000 electrons in each bunch. Clearly there is a limit here; with the same current density, a tube designed for 0·1 mm wavelength would have only *one* electron per bunch! Although the current density could be increased somewhat, there are basic reasons to do with the mutual repulsion of electrons which set a limit to what can be achieved in this direction, so that the 'number of electrons per bunch' criterion is a fundamental one for this type of generator.

Taking everything into account, Convert has estimated that a wavelength as short as 0·1 mm *could* be generated by the backward-wave oscillator; the limit to the shortest wavelength at present generated is set by the current density available rather than manufacture of the comb structure. Whatever the accuracy of this prediction, it is clear that it is correct in order of magnitude, and that the generation of sub-millimetric waves by electron-beam devices will run into greater and greater difficulties as the wavelength approaches 0·1 mm. How closely this wavelength is approached will depend on the extent to which the great technological difficulties can be overcome.

In this section so far we have concentrated primarily on short-wavelength generation as the goal. There are other goals of greater importance in practical application, for example, high power. Very high powers can be generated by the velocity-modulation principle. The generator is a klystron, but not a *reflex* klystron. There are usually *three* resonant cavities, and the electron beam passes through them in succession. In this form the device is an *amplifier* rather than a generator of microwaves. At a wavelength of 10 cm or so it is possible to generate 10–20 MW of power in a short pulse. The contrast with the tens of milliwatts generated by a reflex klystron is striking—klystrons of one kind or another are useful generators of power over a range of a thousand million to one. In the next section we shall look at the application of microwaves to radar, and the usefulness of high-power sources will be apparent in this situation.

Although microwave electron-beam devices have reached an advanced stage of development, recent achievements in microwave semiconductor devices indicate that, at least for low-power sources, vacuum tubes may soon be replaced by solid-state sources. Ever since the invention of the transistor, efforts have been made to extend its frequency of operation to higher and higher values. Transistors are now available for the microwave region, but more recently a number of completely new solid-state microwave sources have been devised, notably the Gunn-effect oscillator and the avalanche diode oscillator.

A full account of either device involves considerable familiarity with solid-state theory, but a simple account of the Gunn-effect oscillator can be given if one basic fact is accepted. In certain semiconducting crystals, such as gallium arsenide, electrons can exist in two different states with different effective masses. Under high electric field conditions a relatively small increase in field causes electrons to change to the heavy mass state. The effect is so marked that the current flow actually *decreases* due to this mass increase when the field is *increased*. This is exactly opposite to Ohm's Law for an ordinary resistance. A resistance absorbs power when an alternating voltage is applied to it, but the gallium arsenide crystal, on the other hand, having *negative* resistance under these special conditions, *generates* power when an alternating voltage exists across it. It can therefore be used as a source of microwave power. This very much over-simplified account neglects many interesting features of the device, but it at least indicates the principal mechanism underlying its operation.

Many new solid-state sources are now being studied, and it is clear that within the next few years solid-state sources will be used for the vast majority of low-power microwave applications.

5. Microwave radar

Much has been written about the use of radar to detect aircraft. This was, of course, the objective of the intensive development of radar during the Second World War. There are, however, a number of other interesting applications of radar, and we shall consider here the use of microwave radar for surveillance from the air. The term 'surveillance' may be taken to include not only mapping in the conventional sense but also the observation of movement of vehicles on the ground. For our purpose, however, it will be sufficient to have the mapping application in mind. To make a good map we must be able to supply enough detail, and this in turn depends on what scale we have in mind for the map. For example, on a scale of 1 inch to a mile a tenth of an inch corresponds to about 500 feet, and clearly we would not be able to *use* information accurate to the nearest foot in such a map, even if we could obtain it. It would be sufficient to work to within a couple of hundred feet for such a map.

Let us now see what limits the detail which a microwave radar can provide. The principle of radar is briefly as follows. A short but powerful pulse of microwave radiation is sent out in a narrow beam in the direction to be examined. A reflecting object (in our case it might be a hill) in the path of the beam returns a very small fraction of the incident energy back to the radar set. The time taken for the pulse to travel to the hill and back is measured and, because the speed of propagation of the pulse is simply the speed of light, the distance of the hill can be found. For example, if the time is 100 micro-seconds the distance is about 19 miles. Suppose now that the duration of the pulse sent out by the radar is one-tenth of a microsecond. As soon as the first part of the pulse is emitted it travels away from the radar, at the speed of light. By the time the last part of the pulse has left the radar the first part will have travelled about 100 feet. Thus the radar pulse travels through space as an electromagnetic signal occupying a distance of 100 feet in the direction of the microwave beam. An object *anywhere within the pulse* will reflect a wave back to the radar, and objects separated by less than 100 feet cannot be distinguished as separate objects. For a 1 inch to the mile map this would not be a serious problem, for, in any case, such objects could not be shown separately on the map. (If we wished to improve the detail to get a larger scale map we would have to use a shorter pulse.) Thus, a pulse length of one-tenth of a microsecond would be quite satisfactory for the one-inch scale we have in mind.

Suppose that the radar beam is pointing east. This calculation we have just performed tells us how much detail our map can have along an east–west line. (The situation is actually a little worse than we have calculated if the aircraft is flying high; we have ignored the difference between slant range and horizontal range in our simple calculations.) We now ask how

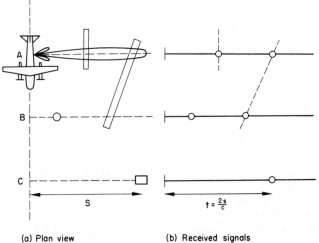

(a) Plan view   (b) Received signals

FIGURE 67 Sideways-looking radar.

much detail we can have in the north–south direction. A north–south line cuts through the radar beam at right angles, and an object on this line anywhere within the radar beam will return a signal to the radar. Two objects so close together that they lie within the radar beam will not be distinguished separately. We see therefore that detail in the north–south direction is limited by the width of the radar beam.

For the case of a radar operating at 3·2 cm wavelength, an aerial length of 15 feet would give a beam width of 0·4°. (This can be verified by calculation from equation (4) in section 3.) At a range of 10 miles the beam width works out to be 370 feet, and this is clearly about right for our 1 inch to the mile map. However, we cannot turn a 15-foot aerial round azimuth to look in different directions; such an aerial is far too large to be rotated in an aircraft. It is at this point that the new principle of the sideways-looking radar is invoked. Figure 67 (a) shows a plan view of an aircraft flying from north to south on a straight line path. It is fitted with a 15-foot-long aerial mounted along its fuselage in a fixed position, and microwave radiation emerges from it in a narrow beam directed eastwards. Although the aerial is long, its height is relatively small, so that the beam is wide in the vertical plane. It therefore has the shape of a fan, and irradiates a strip of ground which may measure ten miles or so in the east–west direction, but only a few hundred feet in the north–south direction. When the aircraft is in position A in Figure 67 (a) the beam intercepts the two objects shown, and pulses reflected from them arrive back at the aircraft after a time delay corresponding to their distance from the aircraft. These pulses are used to brighten the trace of an electron beam which sweeps rapidly across the fluorescent screen of a cathode-ray tube. Two bright spots appear at distances along the trace corresponding to the time delays

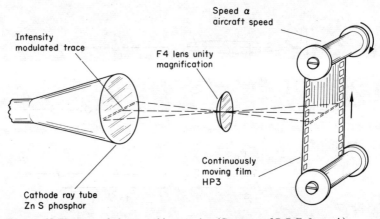

FIGURE 68 Elements of photographic recorder. (Courtesy of R.R.E. Journal.)

of the pulses, and therefore at distances proportional to the actual distances from the aircraft of the reflecting objects on the ground. This is shown in trace A of Figure 67 (b). If a photograph of this trace were taken we would have a one-dimensional 'map' in the east–west direction of the ground over which the aircraft was flying at the instant of taking the photograph.

At a later time the aircraft reaches position B, and again two reflecting objects lie in the beam. Two bright spots therefore appear on the cathode-ray-tube screen at appropriate positions as shown in trace B of Figure 67 (b). Later still, with the aircraft at C, only one reflection occurs, and again this produces a representative bright spot on the cathode-ray-tube screen as shown at C in Figure 67 (b).

If photographs were taken at positions B and C, as well as at A, and the three photographs pieced together properly we would have produced a crude two-dimensional map. It is now easy to see what we should need to do to produce a good map. Clearly, we should take a *very large* number of

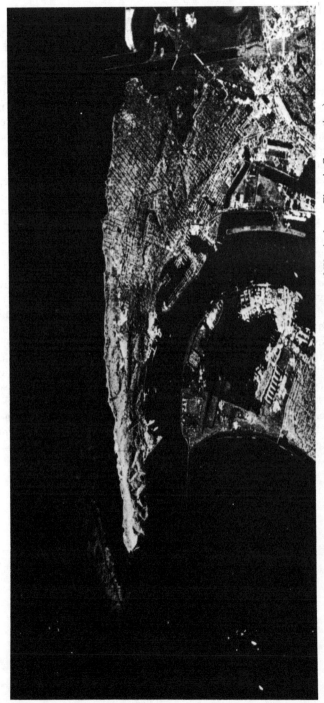

FIGURE 69 Radar map of San Diego harbour, California. (Courtesy of Westinghouse Electric Corporation.)

photographs as the aircraft flies along its north–south line; when these are laid side by side we would have a good map of the region covered. The best way to do this is to photograph the cathode-ray-tube screen continuously on a moving film, as shown in Figure 68. By adjusting the speed of movement of the film in relation to the aircraft speed, the north–south and east–west scales can be made the same, and a conventional map is produced. In a more recent development of the system the lens is replaced by a bunch of glass fibres which transmit the light from the cathode-ray-tube screen directly to the photographic film, thus giving very much improved light transmission.

Sideways-looking radar (S.L.R.) has been developed to a high degree of excellence by the Royal Radar Establishment at Malvern and elsewhere. Figure 69 illustrates what can be done with a modern S.L.R. system. The map of San Diego harbour on the Californian coast shows an impressive amount of detail, and it is clear that, through the development of S.L.R., microwave techniques have made an invaluable contribution to aerial surveillance.

There are, of course, very many other applications of microwaves for both civil and military needs. Satellite communication systems, radar meteorology, air traffic control, detection of moisture content in buildings, microwave ovens and diathermy are a few of the very varied fields in which microwaves have been employed. Apart from these obviously practical applications, microwave techniques have also been employed in pure science, and so have contributed to the advancement of physics and chemistry. Microwave spectroscopy, radio astronomy, gas-discharge physics and particle accelerators have all employed microwaves.

Space, however, does not permit more than this brief mention of the wide and ever-growing range of problems in which microwave techniques are being employed to provide a solution.

SEVEN

CLOSING THE GAP

1. Introduction

In the previous chapters of this book the general properties of electro-magnetic waves have been outlined and explained, and the different regions of the spectrum have also been referred to in some detail. It will now be appreciated that these different regions, stretching from the γ-rays at the very-high-energy and short-wavelength end to the long radio waves of low energy at the other, can be characterised not only by their different wave-lengths but also by the different kind of forces which are disturbed in order to produce the particular radiation.

Thus, high-energy γ-rays are produced when nuclear forces are disturbed and the nucleus itself moves from one energy state to another; X-rays, ultra-violet and visible radiation are all produced by energy changes in the electron orbits, the forces involved being those between the nucleus and its electrons, or between the electrons themselves. The longer-wavelength, lower-energy, infra-red radiation, however, arises from changes in molecular, not atomic, energy states, and the forces involved in this case are those between the atoms themselves, which produce the molecular and solid-state bonding. These are also the forces which are associated with energy changes in the microwave region, which, in turn, merges into the radio regions, where the much smaller energy quanta are usually to be associated with the inter-actions of electrons or nuclei with externally applied electric or magnetic fields, rather than those already existing within the atom or molecule.

The particular energy systems and their associated interactions are summarised schematically in Figure 70, where the forces involved are shown at the top of each region of the electromagnetic spectrum while the practical methods used to study these different regions are given below. In this figure the whole range of wavelengths is depicted as an unbroken band, but in fact it is only in recent years that the production and utilisation of all the different wavelengths throughout the spectrum has been possible. Be-fore the last war generators were in existence for all types of wavelength from the infra-red through the optical, ultra-violet, X-ray and into the γ-ray region, and also, of course, in the ordinary radio-wavelength region, starting from about 1 metre length and extending to wavelengths of hun-dreds of metres. There was, however, a noticeable gap between the long infra-red waves and the shortest radio waves that could be produced. It is this particular gap which is to be discussed in this chapter, and the various ways in which it has now been closed.

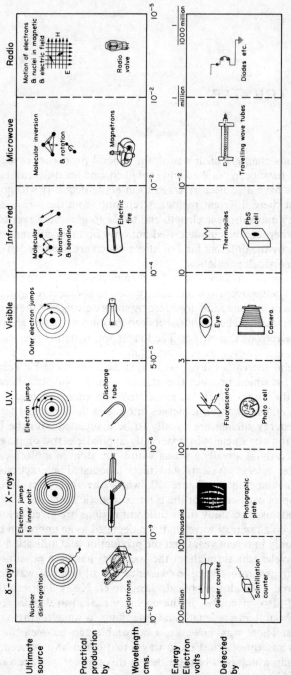

	γ - rays	X - rays	U.V.	Visible	Infra-red	Microwave	Radio
Ultimate source	Nuclear disintegration	Electron jumps to inner orbit	Electron jumps	Outer electron jumps	Molecular Vibration & bending	Molecular inversion & rotation	Motion of electrons & nuclei in magnetic & electric field
Practical production by	Cyclotrons		Discharge tube		Electric fire	Magnetrons	Radio valve
Wavelength cms.	10^{-12}	10^{-9}	10^{-6}	$5\cdot10^{-5}$	10^{-4}	10^{-2}	10^{-2}	10^{-5}
Energy Electron volts	100 million	100 thousand	3		0	$\overline{100}$	$\overline{\text{million}}$	$\overline{1000 \text{ million}}$
Detected by	Geiger counter Scintillation counter	Photographic plate	Fluorescence Photo cell	Eye Camera	Thermopiles PbS cell	Travelling wave tubes	Diodes etc.

FIGURE 70 Schematic diagram of complete electromagnetic spectrum. The ultimate sources of the different types of radiation, indicating the different forces which are involved, are shown along the top of the figure. The practical means of producing and detecting the different kinds of radiation are given below, and the wavelength and energies associated with the different regions are given across the centre.

2. Wartime Work on Radar

The main factor which helped to close this gap in the electromagnetic spectrum was, in fact, all the wartime research that went into radar, with its need for reliable generators of shorter and shorter radio wavelengths. It was this enormous deployment of applied research, and its results, which gave birth to some of the entirely new fields of pure research after the war—i.e. microwave spectroscopy and electron spin resonance. These fields of investigation would never have been possible if this new tool, or technique, had not been so precisely developed in wartime research by the applied engineer. The advances that have come from this interchange between the work of the applied engineer and the investigations of the pure scientist is an excellent example of how these two aspects of science can go along together. The wheel can, in fact, be said to have turned full circle in recent years with the development of the maser and laser, since the basic idea of these came directly from the pure scientists studying resonance spectra, but they are now very much tools in the hands of the microwave and communication engineers.

It will be seen in the sections that follow that the basic concept behind the maser and laser has provided one of the most effective means of 'closing the gap' in our electromagnetic spectrum, since it has allowed coherent radiation to be produced in the shorter-wavelength regions. It would therefore seem to be appropriate that some short account should be given of the way in which these new concepts arose, especially since they themselves came from the work of the pure scientists studying spectra in this new wavelength region.

In the previous chapter, Professor Cullen explained how actual advances in manufacturing techniques and precision of radar valve design allowed the microwave engineers to produce generators of shorter and shorter wavelength. It should also be stressed that these were generators which could be precisely controlled to produce highly monochromatic radiation, as was required for wartime radar purposes. Very fortunately, these were just the properties which could most easily be adapted for pure research in spectroscopy when the war came to an end. A brief historical outline of this development of microwave spectroscopy will serve to show how these two sides of applied microwave engineering and spectroscopy in the centimetre band came to be so intimately connected.

3. Microwave Spectroscopy

The first experiments on microwave spectroscopy at the end of the war were conducted by Bleaney and his co-workers in the Clarendon Laboratory at Oxford, when they measured the inversion spectrum of ammonia gas. The structure of the ammonia molecule consists of a simple pyramid with a nitrogen atom symmetrically placed above a plane of three hydrogen atoms. It is possible for this single nitrogen atom to be in stable equilibrium on either side of the plane of the three hydrogen atoms and, at any normal temperature, the ammonia molecule will have its nitrogen atom vibrating rapidly between these two stable configurations. The frequency of such inversion vibrations is related to the energy jump involved as discussed in Chapter Two, p. 23, and corresponds to a microwave frequency of about 24,000 MHz, or a wavelength of 1.25 cm. It therefore follows that if radiation of this wavelength is passed through a sample of ammonia gas absorption will take place when this critical wavelength is reached, and a precise knowledge of the value of this wavelength will give information on the forces holding the ammonia molecule together.

The essential features of the spectroscope used by Bleaney and his co-workers are shown in Figure 71. The klystron radar valve produces the

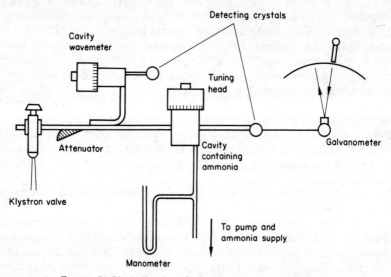

FIGURE 71 Block diagram of microwave gaseous spectroscope.

microwave radiation which is fed to the absorption cell containing the ammonia gas, and the amount of absorption in this sample is measured by the detecting crystal at the end of the run. These three items—radiation source, absorption cell and detector—are, of course, the basic features of any spectrometer working at any wavelength. It is interesting to note, however, that spectrometers in the microwave region do not require any dispersive element, such as a prism or grating, since the source itself can be tuned so accurately that monochromatic radiation is produced.

The spectrum obtained with this simple spectrometer does not consist of one single line corresponding to the frequency of inversion, as this is, in fact, split into a large number of components. These arise from the rotation of the ammonia molecules which can occur in a variety of different ways at the same time as the molecules are undergoing inversion. A study of these second-order splittings will therefore give information on the rotational

energies and interactions of the molecule. This is not the only extra information that can be obtained. Soon after Bleaney and his co-workers had
published their results Good of Westinghouse Laboratories in America
showed that much greater resolution could be obtained by working at low
pressures and low power levels, and was able to resolve a far finer spectral
structure, known as a hyperfine structure. This arises from an interaction
between the quadrupole moment of the nitrogen nucleus and the molecular
field of the ammonia molecule. This quadrupole moment measures the
distortion of the nitrogen nucleus from spherical symmetry, and so we find
that a study of second- or third-order splittings on the main spectral line
can give very interesting and important additional information—in this
case concerning the structure of the nucleus itself. The very high order of
resolution available at these wavelengths, and thus the ability of microwave
spectroscopy to probe into the depths of atomic structure and obtain very
precise information on nuclear–electron interactions, has been one of its
most striking successes.

4. Electron Resonance

While this new technique was being extended to study a wide variety of
different molecular gases, another field of research was also opening up
which employed microwave radiation. This time, however, the microwave
radiation was used to probe the energy states of electrons in a solid rather
than those of molecules in a gas. This new line of research is called electron
spin resonance, and a large magnetic field is normally employed so that the
energy levels of the electrons to be studied have differences which fall in
the microwave region.

The general principle underlying this technique is illustrated in Figure
72. This is drawn for the simple case of a solid containing free unpaired
electrons (see p. 34) which interact with the applied magnetic field. In
the absence of such a field, all the electrons will have the same energy,
whatever the orientation of their spins and associated magnetic moments.
If, however, an external magnetic field is applied across the specimen
the electron spins and magnetic moments will align themselves so that they
are either parallel or anti-parallel to the direction of the field (no other
orientation being permissible by quantum conditions). Those aligned anti-
parallel to the field will have more energy than those aligned parallel, as
can be seen by analogy with the unstable and stable positions of a simple
bar magnet in a magnetic field. The electrons are therefore split into two
energy groups, as shown in the figure, the difference between these being
given by $g\beta H$, where β is the Bohr magneton, the smallest unit of magnetism allowed by quantum theory, and g is a factor which effectively
measures the amount of orbital and spin momentum possessed by the
unpaired electron, and has a value of about 2·0 for a free electron
possessing no orbital motion.

The essential feature of an electron resonance experiment is to separate
the electrons into two such groups by the application of an external

magnetic field, and then to place the sample in a region of electromagnetic radiation of frequency, ν, such that

$$h\nu = g\beta H \qquad (1)$$

The incident photons are then of just the right energy to excite the electrons from the lower to the upper energy level, reversing their spins and moments in the process. Substitution of numerical values into this equation gives a resonance condition of

$$\nu = 1\cdot4 \, . \, 10^6 \, . \, gH \quad \text{Hz} \qquad (2)$$

and it can be seen that if the applied field is 10 gauss the resonance frequency will be about 30 MHz, whereas if the applied field is 10,000 gauss the resonance field will be about 30,000 MHz.

There is no reason, in principle, why electron resonance should not be performed at any frequency, provided that the corresponding magnetic field is applied, but there is one important practical reason why most measurements are made at as high a magnetic field strength and frequency as possible. This concerns the intensity of the observed resonance signal, and hence the sensitivity of the apparatus. In the same way that electromagnetic power is absorbed in raising electrons from the lower to the upper level, so it is emitted when electrons fall from the upper to the lower level. The coefficients of absorption and stimulated emission are equal, and the only reason why any net absorption signal is obtained is because there are more electrons in the lower than in the upper level.

It follows that a larger absorption signal will be obtained if there is a larger difference in population between the two levels, this difference in population increasing exponentially with the actual difference in energy between the two levels. So greater absorption, and thus greater sensitivity, will always be obtained by working with larger applied magnetic fields and higher applied frequencies—hence the use of microwave radiation for this form of spectroscopy.

There is another important concept that follows from a consideration of the simple energy levels shown in Figure 72. This concerns the 'relaxation' of the spins back to their normal population distribution. If the process of absorption and stimulated emission were the only means whereby the electron spins could exchange energy the net absorption would rapidly cease since the electromagnetic radiation would very quickly equalise the numbers in the two energy levels. This, however, would occur only in a completely isolated spin system and, in practice, the spins are also coupled to the thermal vibrations of the whole solid and can lose extra energy by sharing with these. This type of interaction is termed 'spin-lattice' interaction, and it has the effect of trying to restore the normal population distribution. If this spin-lattice interaction is strong the populations of the levels will be given by that corresponding to normal thermal equilibrium, whatever the power level of the incident electromagnetic radiation. If the spin-lattice interaction is weak, however, the difference between the populations of the two levels will be steadily decreased as the incident power level is raised, and there will be a reduction in the intensity of absorption, as compared with the first case. This phenomenon is known as 'saturation' of the energy levels by the microwave power, and will be seen to play a very important role in the action of maser amplifiers and oscillators. It may be noted that interaction with the lattice always becomes weaker as the temperature is decreased, and 'saturation effects' will therefore be most pronounced at the lower temperatures.

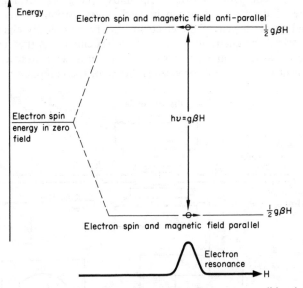

FIGURE 72 Basic principle of electron spin resonance. Resonance condition is given by

$$h\nu = g\beta H$$

I.e. $\nu = 2 \cdot 8 \times 10^6 \times H$ for free electrons. Thus at 10,000 gauss $\nu = 28{,}000$ MHz.

The kind of apparatus which is required to investigate 'electron resonance' is very similar to that employed for the gaseous microwave spectrometers, but a magnetic field is now added to produce the required energy level splitting. The simplest type of spectrometer is therefore as shown in Figure 73. The klystron (see p. 115) feeds the radiation along the waveguide run to the cavity resonator, which acts as the absorption cell. In this way the sample can be placed in a highly concentrated microwave field, and the cavity containing the sample can be inserted between the pole pieces of an electromagnet. In the transmission system shown a second waveguide then leads from the cavity to a crystal detector, where any change in signal level produced by absorption in the cavity is passed on to the display system. The resonance absorption is obtained by varying the strength of the magnetic field until a dip occurs in the detected microwave power. If a small a.c. magnetic field modulation is also applied the resonance absorption will be swept through twice in a cycle, and may be amplified and displayed on an oscilloscope as shown. Measurements may be carried out at low temperatures by placing the cavity in a vacuum flask containing liquid oxygen, nitrogen, hydrogen or helium, the whole assembly being inserted into the magnet gap.

This kind of apparatus was employed by physicists working in University laboratories after the last war to study the properties of electrons in the solid state and, in particular, the energy levels associated with them in different paramagnetic compounds. The great advantage of this technique of electron resonance in such investigations is the fact that individual energy levels of the electrons can be completely characterised, and the overall magnetic properties of the specimen can then be deduced from these, rather than vice versa, as is the case with measurements on specific heat or other bulk properties. Some of the compounds studied in

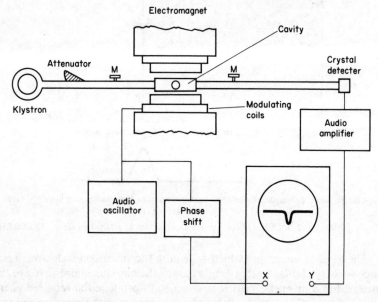

FIGURE 73 Block diagram of simple crystal-video electron resonance spectrometer. The absorption is shown displayed on the oscilloscope screen, while the timebase is swept in synchronism with the magnetic field modulation.

this way were the transition group ions, such as iron, manganese and chromium. These investigations were first made just to obtain a better understanding of the energy level systems in such compounds, and hence more detailed information which would be of use for general theories of atomic structure and magnetism.

It was out of such pure scientific research as this that the idea of the maser was born, and this particular development is a very good example of the way in which investigations which are being conducted for the sake of pure research can suddenly produce a new line of immense interest in applied science. It might now be appropriate, therefore, to leave the particular topic of electron resonance, and to outline briefly the way in which the concept of the 'maser' arose from these studies.

5. Masers

When discussing the general principle of electron resonance in the last section it was pointed out that the only reason why a net absorption of radiation occurred was because there were more electrons in the lower than in the upper energy level. This condition of a larger number of electrons or atoms in the lower energy level always exists in any system which is in thermal equilibrium. The difference in population between the two levels is, in fact, determined by the competition between the thermal energy, which is tending to randomise the electron distribution between the levels, and the energy level splitting, which tends to drain the electrons into the lower level. By using Statistical Mechanics (see p. 23) one can express the actual ratio of the number in the upper level, n_1, to the number in the lower level, n_2, by the expression

$$\frac{n_1}{n_2} = \exp\ \left(-\frac{hv}{kT}\right) \tag{3}$$

This is known as the Maxwell–Boltzmann distribution, and applies to the vast majority of physical systems under ordinary conditions of temperature. The energy level splitting between the two levels is given by hv, and it is this splitting which is effectively in competition with the thermal energy given by kT, where k is Boltzmann's constant and T is the absolute temperature. The negative sign associated with the expression in the bracket shows that the number in the upper level, n_1, is always less than n_2 under conditions of thermal equilibrium, and it is for this reason that absorption, rather than emission, of the microwave radiation takes place in an ordinary electron resonance experiment.

It follows, however, that if, by some artificial means, it is possible to invert this population distribution so that there are more electrons in the upper level than in the lower, then microwave emission instead of absorption will be obtained. It is this artificial inversion of the populations of the energy levels that is the essential idea behind the maser. This acronym stands for 'Microwave Amplification by Stimulated Emission of Radiation', and it is a device which employs the principles of electron resonance in reverse so that microwaves are emitted (or amplified) instead of being resonantly absorbed (see p. 35). One extremely important aspect of stimulated emission is that the waves emitted by different atoms are exactly in step. This is quite different from the usual case, where the photons, and hence the waves, are emitted randomly, and consequently are not in step. We shall return to this point later when discussing the *coherence* of the radiation. There is a variety of methods whereby the artificial inversion of the energy level populations can be obtained, some requiring pulse operation and rapid change of field strengths. The most promising method, however, produces a continuously inverted population system by a relatively simple method, and is designated the 'three-level maser'. This was first suggested by Bloembergen in the U.S.A., and Basov and Prokhorov in Russia, and the method affords a very good example of general maser principles.

Before this method of operation is discussed in detail the concept of the 'electronic splitting' of an electron resonance absorption must be introduced. So far we have only considered atoms containing one unpaired electron, but many transition group atoms contain two, three or more unpaired electrons. The total spin and magnetic moment can be found by adding the individual components. Since each electron has a spin of $\frac{1}{2}$ in quantum theory, the total spin S of three unpaired electrons would be $S = 3/2$ if all the spins were directed along the external magnetic field, or $S = -3/2$ if they pointed in the opposite direction. If we now consider the particular case of an atom with two unpaired spins, then the net spin will be able to align itself in three ways with respect to an externally applied magnetic field. These may be denoted $+1$ (with spins parallel to field), 0 (spins perpendicular to field) and -1 (spins anti-parallel to field). The energy of the atom will depend upon the arrangement of the spins (and, of course, upon the strength of the external magnetic field), so that three slightly different energies are obtained corresponding to the different arrangements as shown in Figure 74. Only when the external field is reduced to zero does the energy gap between three levels disappear. Because the energy difference between the $+1$ and 0 states is the same as that between the 0 and -1 states, then both resonance transitions will be induced by the same radiation, $h\nu$. In a solid, however, there are often strong internal electric fields due to the arrangement of the crystal. Under these conditions the energy of the 0 state is slightly different from that of the -1 and $+1$ states, even when there is *no* external magnetic field, and the energy levels are as shown in Figure 74 (b). In this case the energy gap between the $+1$ and 0 states and the 0 and -1 states is not quite the same when an external magnetic field is applied. Incident radiation, $h\nu$, will therefore induce resonance transitions $+1$ to 0 and 0 to -1 at two slightly different field strengths H_1 and H_2, and hence two absorption lines are obtained. This is the phenomenon called 'electronic splitting'.

The electronic splitting produced in this way reflects the symmetry and strength of the internal crystalline electric fields and can, of course, give very useful information on the solid-state properties. Since there is, in effect, a competition between the internal crystalline electric field, the direction of which is determined by the orientation of the atoms in the crystal, and the effect of the externally applied magnetic field, it follows that the magnitude of this electronic splitting normally varies quite markedly with the angle between the applied magnetic field and the crystalline axis. We shall see later that this is an extremely important and useful fact,

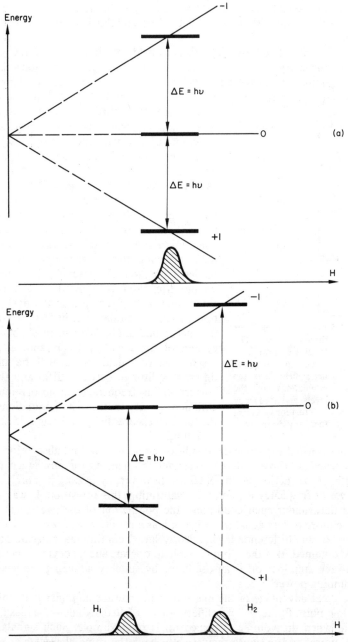

FIGURE 74 Splitting of energy levels for two coupled electron spins, i.e. $S = 1$.
(a) With no zero-field splitting.
(b) With a zero-field splitting due to the internal crystalline electric field.

since it does enable the energy-level splittings to be altered by simply rotating the crystal within the magnetic field, and this in turn enables solid-state masers to be tuned.

Another way of representing this zero-field splitting is shown in Figure 75. Here the energy level positions of the three states are shown for a given value of applied magnetic field. The transitions between these are labelled ν_{12}, ν_{23}, and ν_{13} respectively as shown. The one for ν_{13} is drawn with a dashed line, since this is normally a transition which is forbidden under a series of selection rules which state that the resolved components of spin must only change by unity for any normal transition—not by two as in this case. There are, however, second-order factors which can be of quite large magnitude in solid-state systems, and these terms tend to blur the otherwise precise quantum assignment of the energy levels, thus removing the restriction on the forbidden transition. In practice, therefore, a double transition is allowed to a certain extent and, as will be seen in the next sections, such transitions are, in fact, essential for maser operation. We can thus obtain a system of three energy levels and, at any given magnetic field, there will be three different splittings between them and three corresponding frequencies of absorption or emission, as designated. The populations of the three levels are shown in Figure 75 as n_1, n_2 and n_3.

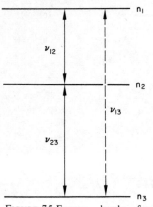

FIGURE 75 Energy levels for three-level maser. The levels are shown for a given applied magnetic field and the populations of the levels and frequency of transitions between them are designated.

Under normal experimental conditions $n_1 < n_2 < n_3$ and absorption can be observed at three different resonant frequencies for any given field strength. Maser action can be obtained, however, by feeding in a high level of power at frequency ν_{13} and thus 'saturating' this transition. It was seen earlier that, under such conditions, the population of the two levels will become more or less equal ($n_1 = n_3$) and, as a result, n_1 will become greater than n_2. If this difference is sufficiently large, continuous emission at ν_{12} will be obtained. It is therefore possible to operate such a device either as a microwave amplifier or as an oscillator by suitably adjusting the level of the pumping power.

The great advantage of the maser as a microwave amplifier is its inherently low noise figure. All amplifiers add unwanted signals—the hiss from a radio when the volume is turned up is one example. Such signals are called 'noise' and, because they usually have the same character as electrical signals associated with heat radiation, their strength is conveniently expressed as a temperature in degrees Kelvin. No semi-conductor contacts involving the large excess flicker noise of silicon detectors are present, and neither are there any hot electron beams, as in normal valve devices. The noise arises solely from the effective resistance at the temperature of operation and, since this is usually 4° K, the inherent noise figure of the amplifier can be made extremely small: in masers employed in such projects as the transatlantic television links an overall noise temperature of only 10° K can be obtained, whereas conventional amplifiers have a noise temperature of over 1,000° K.

It can thus be seen that there are five essential requirements for any maser of this three-level type. First there is the choice and provision of the

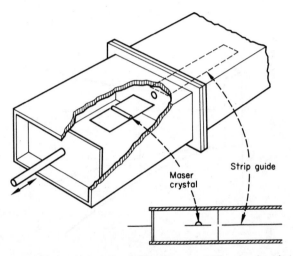

FIGURE 76 Diagrammatic view of first solid-state maser. The signal to be amplified is transmitted along the strip-guide, while the pumping power is supplied along the rectangular waveguide to the cavity.

basic paramagnetic material which must contain such ions as trivalent chromium in low concentration. Secondly, the application of a large magnetic field is required to split the energy levels of the ions in the way indicated in Figure 75. The actual spacing of these energy levels has, of course, to be chosen so that two of the transitions to which they give rise correspond to the frequency to be amplified and the pumping frequency. The third basic requirement is the provision of the pumping frequency itself, this being the highest of the three frequencies indicated in Figure 75; high power will be required to saturate this transition. The last two basic requirements are of a practical nature. An input and output waveguide system or co-axial line is required to feed the incoming signal to the maser material, and then to accept the amplified radiation and feed it to a second-stage amplifier. Finally, a low-temperature system employing liquid helium will be required to cool the whole device to 4° K.

The first maser actually to operate in this way was produced by Scovil and his collaborators at the Bell Telephone Laboratories, and a diagrammatic view of this device is shown in Figure 76. A crystal containing rare-earth ions was employed, and this crystal is shown mounted in the centre of the device on a strip inside the microwave cavity. The higher-frequency pumping power is fed down the waveguide to the cavity itself, while the strip waveguide, which forms a transmission line for the lower frequency to be amplified, is continued into the cavity so that this is also resonant

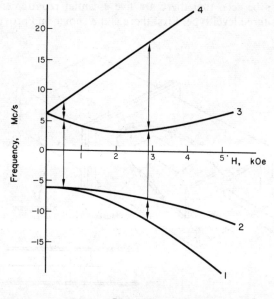

Energy levels ruby θ = 90°

FIGURE 77 Energy levels for the ruby maser. The variation of the four energy levels of the chromium ions as the applied magnetic field is altered is shown, together with the possible transitions that can be employed for maser operation.

at the longer wavelength. This dual-frequency resonant cavity is then immersed in liquid helium and the Dewar flask placed between the pole pieces of a strong electromagnet. Following the successful operation of this first solid-state maser, others have been developed using different paramagnetic ions as the active material.

It has been found that ruby, which is aluminium oxide with a small percentage of chromium-ion impurity, is one of the best of such active materials. The maser in the Goonhilly Downs receiving station for the transatlantic television link has ruby as its active element, and its energy levels are shown in Figure 77. The microwave side of the device has also been developed since the first measurements and, instead of using a single crystal of ruby, modern masers employ a travelling-wave type of structure.

This arrangement is shown in Figure 78, where a comb-type metal structure is being used to slow the microwaves by a factor of about 100 (see p. 117) and to transmit oscillations with a range of frequencies of about 370 MHz. The ruby itself is mounted along one side of the comb structure and the incoming signal is continuously amplified as the wave traverses this active material.

This particular maser has been described in a fair amount of detail, since it is typical of those that are now in everyday use in conjunction with a very important practical application. It is probably fair to say that, even in terms of today's rate of development of science, it is very rare to find a fundamentally new idea exploited as quickly as this initial suggestion of Bloembergen that amplification of microwave radiation should be possible in a device such as this. It is extremely gratifying to see this major contribution by the pure physicist to the production of a really important device for communication purposes, especially when one considers how much the whole field of magnetic resonance research itself owes to the work of the radar engineers in the first place.

At the moment it may appear that we have strayed rather a long way away from the main theme of this particular chapter, which was, as you may need reminding, the idea of closing the gap in the electromagnetic spectrum. It is quite true that masers themselves operate in a frequency region for which other generating devices, such as the magnetron or klystron, are already known. Masers themselves do not, therefore, make any direct contribution to the closing of the frequency gap in the electromagnetic spectrum, but they do (like other microwave generators) produce coherent radiation; and the great importance of the concept of the maser is that this basic principle can be rapidly extended to much higher frequencies, where all other attempts have failed. In this way the concept of the maser merges into that of the laser, which is basically the same device, operating by an inversion of energy-level population, but which produces emission in the optical instead of the microwave region.

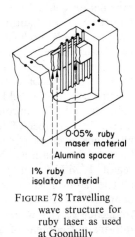

0·05% ruby maser material
Alumina spacer
1% ruby isolator material

FIGURE 78 Travelling wave structure for ruby laser as used at Goonhilly Downs.

6. Coherent and Incoherent Radiation

Before considering the principles and construction of lasers themselves it will probably be wise if a few moments are spent stressing the difference between coherent and incoherent radiation. The former consists of a continuous succession of regularly spaced waves (as shown in Chapter 1, Figure 9), which are all in phase, i.e. with maxima and minima coincident. Radio and microwave engineers are very familiar with this concept, since all the carrier waves used in communication are examples of coherent radiation. It is possible to modulate such carrier waves with complex amplitude and frequency patterns, the details of which will be retained on the carrier wave when it is demodulated, and it is this particular feature which enables more than one channel of information to be carried on a single carrier wave. If, however, the carrier wave were made up of wave trains which lacked any phase coherence between them (such as that shown in Chapter 3, Figure 22), then only a very gross form of amplitude modulation could be effected by altering the overall intensity of the incoherent beam. In addition, it is not possible to transmit a large number of such amplitude-modulated signals on a single carrier frequency, as is the case with coherent radiation. It is for this reason that, until very recently, no attempt had been made to use visible radiation itself as a carrier wave for the transmission of television or other signals.

However, in the field of communications the possibilities that arise if coherent radiation becomes available in the visible region are quite enormous. The number of television programmes that could be transmitted along one given carrier wave may serve as a useful illustration of the quite fantastic implications that arise from coherent radiation in the visible region. It will be well known to any who have any acquaintance with radio or television broadcasting that it is necessary to impress the information contained in the actual television programme on to a much higher-frequency carrier wave before this is broadcast. Thus, although the actual information content of a television programme occupies a bandwidth of only about 3 MHz, the frequencies of the carrier waves which are actually broadcast to carry this signal are greater by a factor of 10 at about 50 or 100 MHz. Moreover, if several such television programmes are to be broadcast at once along the same transmitted beam, then it is an advantage to use much higher-frequency carrier waves still. This is one of the reasons why microwave links working in the thousands or tens of thousands of MHz frequency region are so extensively employed in television transmission. In fact, a coherent beam of microwave radiation of 4,000 MHz can carry about one hundred times the information that can be transmitted in a coherent beam at 40 MHz.

We can now see what will happen if this idea is extended into the visible region. There would, for example, be a bandwidth of 3×10^7 MHz between radiation of a wavelength of 4,000 Å and that of wavelength 7,000 Å, and it would be possible to transmit simultaneously all the information required for over a million television programmes on such a light beam of coherent radiation. It is the enormous potential information capacities of coherent beams at these frequencies which is one of their fascinations, and promise to be one of their main applications.

It can now be seen that the possibility of producing coherent radiation in the infra-red and visible region by an exactly similar process to that employed in the microwave region has effectively closed the gap, in so far as different types of radiation are concerned. The advent of the 'laser' and its infra-red companion the 'iraser' following on from the invention and construction of the 'maser' can thus be considered as linking together the different regions of the electromagnetic spectrum in a very fundamental way—it is now possible to produce coherent radiation in all these different wavelength regions by precisely controlled atomic processes rather than by the random thermal excitation that has been the only method available in the past.

7. The Laser

The general principles and ideas behind the construction of the laser are identical to those of the maser. The first two general requirements can be summarised as, firstly, an available working medium with a suitable energy level system. This must contain two levels with an energy difference corresponding to the frequency of desired operation, and must also contain other levels above these so that saturation pumping or similar effects can invert the energy population of the first two. Secondly, some means must be devised of producing this inverted energy level population so that there is a higher probability of stimulated emission than of resonance absorption. The third general requirement is a quantitative one, and this is that a sufficiently large number of atoms must be in the upper level so that the resultant stimulated emission can overcome circuit losses and similar effects. This requirement is met in microwave masers by the use of a cavity resonator which, effectively, reflects the microwaves to and fro many times, thus enabling them to interact with a large number of atoms in the upper state. The system for lasers analogous to this was initially suggested by Towns and Schawlow. They thought that sufficient path length could be obtained by using a reflecting system very similar to that employed in a normal Fabry–Perot etalon. In this system two mirrors are aligned accurately parallel to one another so that a ray of light travelling normal to the plane of the mirrors will be reflected to and fro between them a large number of times. Hence, if the gain due to excess stimulated emission in the active medium, which is placed between the mirrors, is greater than the losses which occur on reflection at the mirrors there will be a net amplification of radiation during each passage; in this way a large amplitude of radiation of a particular frequency will be built up in the system. It follows that, if a suitably activated medium can be inserted between the mirrors, coherent monochromatic radiation with highly directional properties should be produced.

It might be helpful if we now consider the actual development of experimental lasers since the first one, which was designed and constructed by T. H. Maiman of the Hughes Aircraft Company, operated in July 1960. He employed a ruby crystal as the activated medium, and the reflecting surfaces were obtained by polishing the two ends of this crystal very accurately so that they were parallel to better than 6 seconds of arc. If a solid crystal is used as the active medium the demands on the homogeneity of the optical properties of the crystal itself are also very great. This is one of their main limitations. A diagram of the ruby laser as designed by Maiman is shown in Figure 79. It can be seen that the back end of the crystal is completely silvered, whereas the front end is only partially silvered, and it is through this end that the pulse of coherent radiation is emitted when laser action takes place.

The active medium in this laser was formed by chromium atoms, as was also the case in the maser; in addition, both devices used ruby (aluminium oxide) as the host material. With the laser we are concerned with much larger transitions within the energy level system of the ruby, and the energy levels associated with the laser action are shown in Figure 80; the most important of these is level C, which is metastable. If the chromium atoms are excited into this level they will have a lifetime of about one-hundredth of a second unless stimulated to emit beforehand. The radiation

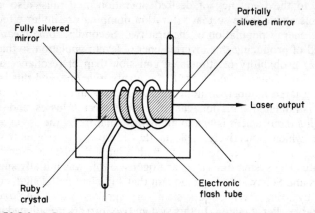

FIGURE 79 Schematic diagram of first ruby laser. The polished and silvered ruby crystal is shown surrounded by the coil of the electronic flash tube.

emitted when atoms return from this level to the ground state has a wavelength of 6,943 Å, and constitutes the red fluorescence of ruby crystals. It was at this particular wavelength that laser action was first produced.

We have seen that the essential requirement for obtaining such coherent radiation is to produce a larger population of chromium atoms in this higher metastable state than in the ground state. The chromium atoms can be raised to this metastable level by pumping them to the higher levels, A or B, and then letting them decay to level C, which they will do relatively quickly. All that is required, therefore, is a source of radiation with a frequency range which will cover the transition from the ground state to level A, and also of sufficient power to excite a large number of chromium atoms to the upper levels. The situation will then be as indicated in Figure 80 (b). These atoms will then quickly fall to the metastable level C and be available for laser action when stimulated by incoming radiation of the correct frequency. Maiman's particular contribution was the discovery that an electronic flash tube could supply all the required pumping power and that this could be wound around the ruby crystal as shown in Figure 79.

The large amount of internal heating which occurs when a flash tube is used means that this system could only be employed to produce pulses of coherent radiation, and could not act as a continuous source. It was found that the power level of the electronic flash tube was quite critical; below a certain level only the incoherent red phosphorescence of the ruby, spread over its normal wide range of frequencies, was obtained. Once the level of the electronic flash was increased above the critical value, an intense beam of red light would be emitted within a period of 10^{-4} second after the

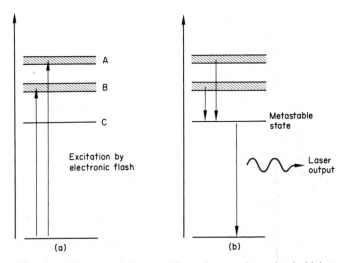

FIGURE 80 Energy levels of chromium atoms in a ruby crystal associated with laser action.
    (a) Before pumping.
    (b) After pumping.

electronic flash. This indicated that sufficient chromium atoms had been pumped up to the excited state to ensure that the stimulated emission they produced was greater than the losses which occurred at the ends of the optical system. This practical demonstration that coherent optical radiation could be produced via a process of stimulated emission confirmed the basic ideas and theories concerning laser action. Quite a large number of other lasers operating with similar single crystals have now been built and operated, but all these suffer from two inherent drawbacks. First, since the activated medium is in the solid state, there is a definite limit to its homogeneity, and therefore to the directional coherence of the emitted radiation. Secondly, the thermal stresses set up in the solid by the absorption of radiation from the electronic flash tube prevent continuous operation, and hence the use of efficient modulation techniques.

The two main disadvantages of the solid-state lasers can be overcome if a gaseous medium is used to supply the activated atoms. The first such laser employing gas molecules was operated successfully by Javan of the Bell Laboratories in 1961. This could be made to produce a continuous output of coherent radiation and required only 50 watts of input power. The basic ideas were very similar to the ruby laser, in that a Fabry etalon reflecting system was again employed, and the two mirrors were mounted at either end of a glass tube containing the gas.

The activated medium consisted of a mixture of neon and helium gas, and the energy level of this system is shown in Figure 81. (The notation 2S, 2P, etc., refers to particular excited states of these atoms. A detailed account of its meaning is not possible without a considerable extension of

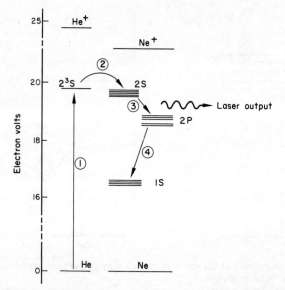

FIGURE 81 Energy levels of helium and neon atoms as used in first gaseous laser.

the simple treatment given in Chapter Two.) The crucial level in this case is the 2S level at the top of the neon system, laser action being produced by stimulated emission, which takes place between this level and the next, the 2P level. The necessary large number of excited atoms in this 2S level is obtained by an indirect method employing the presence of helium gas. Helium has a long-lived metastable state, i.e. 2^3S on the left of Figure 81, and atoms in the ground state of the helium atom can be excited to this metastable state by such processes as impact in a gas discharge. Since the excited helium atoms will then be in the same energy state as that corresponding to the 2S level of the neon atoms, it is very simple for them to transfer their energy by direct collisions with neon atoms, exciting them directly to the top 2S level. In a suitable mixture of helium and neon gas it is therefore possible to build up a dynamic equilibrium in which there is a large number of atoms in the 2S level of neon, although this is not a metastable state itself. It may be noted that two features distinguish this energy level system from that used in the ruby laser. First, the excess population in the higher level associated with the laser transition is now achieved not because of the particular properties of this level itself but because it can be fed from a level of equal energy in another atom with metastable properties. The second feature is that the pumping energy is supplied in this case not directly by a photon of higher energy, as in the electronic flash tube, but by collision with an accelerated electron in a gas discharge. It is possible in principle, of course, to use any method of excitation.

A block diagram of this gaseous laser is shown in Figure 82. The main chamber containing the helium/neon mixture consists of a quartz tube 80

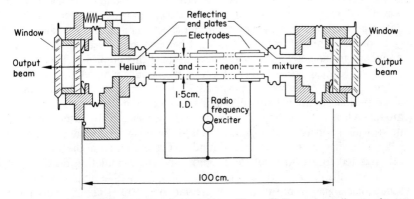

FIGURE 82 Block diagram of first gaseous laser. The radiofrequency coils are shown surrounding the main tube and the reflecting mirrors are held by the bellows at each end and can be adjusted as shown.

cm long and of inside diameter 1·5 cm, which is terminated at each end by the larger metal chambers holding the Fabry–Perot reflecting plates. These can be adjusted by flexible bellows, as shown. The separation of the plates themselves is 1 metre, and the discharge is excited by external electrodes fed from a 28-MHz generator. Laser oscillations were observed from this device at five different wavelengths spread through the infra-red region, and continuous output of coherent radiation was thus obtained with powers of the order of 15 mW. Since the operation of this first gaseous laser, a large number of laboratories in different countries have pressed forward with this work, and quite a variety of different materials have now been employed as the active medium. As a result, masers have now been reported operating at a large number of different wavelengths through the infra-red and visible regions.

It was demonstrated fairly soon after the initial gaseous laser itself had been developed that such lasers could be employed for amplification as well as for the generation of optical radiation. Both of the lasers described so far are actually light oscillators rather than light amplifiers, as in each case the whole process of stimulated emission is initiated by a noise signal present in the active medium, and the gain and feedback of the system are sufficiently large for such a noise signal to produce oscillations. In a large number of applications, however, it is much more useful if the device will act as an amplifier rather than as an oscillator. The first successfully designed and constructed coherent light amplifier was therefore a significant step forward. This was achieved by Jacobs, Gould and Robinowitz, who employed caesium vapour as the active medium, this vapour being excited by selective optical pumping following a method suggested earlier by Schawlow and Towns. In this method intense radiation from a helium discharge lamp is used to excite the caesium atoms to an upper energy level, the population of which can thus be kept larger than either of two lower levels. It is thus possible to obtain laser action at a wavelength of 3·2 microns, which is in a wavelength region where sensitive detection methods employing lead sulphide cells are readily available.

In order to prove that coherent light beams could be amplified, it was necessary to have a source of radiation to feed this amplifier. This source was also formed from activated caesium vapour pumped by the same helium line—in other words, it was basically the same device, but operated as an oscillator rather than an amplifier. The output from this source was modulated at a frequency of 105 Hz and then fed to the laser amplifier itself, which consisted of a caesium cell 90 cm long, illuminated by a long helium discharge lamp mounted behind it. The output from the amplifying cell was fed to a lead sulphide detector, and the modulation on the light beam was displayed and recorded in a phase-sensitive amplifier. It was found that signal-to-noise ratios of greater than 100–1 could be obtained from this laser amplifier, and its amplifying action was confirmed by switching off the various activating elements in turn. The experimentally observed gain was of the order of 6 per cent for the 90-cm cell length, and although this was not as good as predicted by a theoretical analysis, the fact that amplification of the light could be produced in this way did complete the second essential step in the realisation of laser techniques.

The next significant advance in the development of lasers came with the discovery that energy level populations could be inverted in ordinary semiconductor type material. Semiconductors are an extremely important class of materials with properties between those of metals and insulators. In such materials the application of heat excites electrons to higher energy levels, in which they are free to move, hence enabling an electric current to flow as in a metal; alternatively, the vacant space or 'hole' left by an electron when it moves to a higher energy level may behave like a positively charged electron, which again allows an electric current to pass. Using semiconductors, it is possible to produce a continuous output of coherent radiation. The frequency of the emitted radiation corresponds to the energy jump which is involved when an electron falls back into a 'hole' in the semiconductor medium.

There are, of course, other means whereby electrons and holes can lose their energy and not produce radiation in the process. Other such interactions include collisions with the crystal lattice or with other electrons, holes or lattice defects. In certain cases, however, it is found that the probability of electron–hole recombination occurring is much higher than the probability of the other interactions taking place.

The first such devices were those employing gallium arsenide, and the general features of this type of semiconductor laser are outlined in Figure 83. It is seen to consist of a crystal of gallium arsenide which has had its opposite faces accurately polished so as to produce the normal Fabry–Perot type of reflection system. After being reflected many times the radiation emerges at right angles to the polished face as shown, and observation of the interference fringes produced by such radiation indicates that a high degree of coherence over the whole of the emitting region exists.

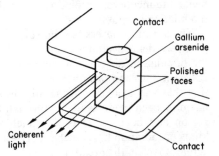

FIGURE 83 Block diagram of semiconducting laser. The gallium arsenide crystal has polished faces as shown, and the beam of coherent radiation is emitted normal to these surfaces.

The great advantage of these lasers compared with those that have been previously described is that there is a *direct* conversion of electrical energy into visible radiation, and hence it is an extremely simple matter to modulate the output radiation. Semiconductor lasers have developed extremely rapidly since these first observations, and they have now been constructed to operate not only as emitters of coherent radiation but also as amplifiers. Since the whole field of laser development is changing so fast, it would not be useful to summarise more recent devices in any detail, since these will rapidly be superseded by newer and more efficient versions. From this brief description it will be evident, however, that the laser has now become a practical solid-state device which can be used both as a source of coherent oscillations that can be easily modulated and also as an amplifying device in its own right.

8. The Gap is Closed

This chapter has been very much concerned with a description of such devices as masers and lasers, because not only have these devices closed an actual gap in wavelength in the electromagnetic spectrum, but they have also provided a method of obtaining a quite different type of radiation. Thus, before the advent of the maser and laser, coherent electromagnetic radiation suitable for detailed transmission of information was being generated easily across the whole of the radiofrequency region and well up into the microwave region itself. On the other hand, although sources were available in the infra-red and optical regions, none of these could produce coherent beams of radiation. In all cases they relied on some form of random excitation, either by simple thermal vibrations of a heated solid or from the collision processes in a gas discharge. It was in this entirely uncoordinated manner that atoms or molecules were excited to the higher energy state and from which they then fell, emitting the required radiation. There were no means then available of aligning the phases of these emitted photons or of persuading them to radiate in step with one another, and hence the incoherent radiation so produced was of no use for any precise modulation purposes.

The advent of lasers has, however, completely altered this picture, since precisely defined atomic-energy-level systems can now be employed and the population distribution between the energy levels be inverted so that coherent emission of radiation is obtained. In this way, the coherent radiation that was available in the microwave region and at lower frequencies has now been made available through the infra-red region, into the visible region and beyond. Thus the much more fundamental gap of different kinds of radiation has now been closed, as well as the more practical one of actual wavelength spread.

Possibly, however, the most interesting aspect of the research summarised in this chapter is not so much the actual closing of the wavelength gap but the way in which advances in pure science and engineering are inter-related. It was pointed out earlier that microwave spectroscopy and electron resonance, which have both been such fruitful areas of research for the pure scientist since the war, only became practicable fields of study because of the rapid advance in microwave technique that was made as a result of the wartime research on radar.

In this way the pure scientists were handed a tool by the microwave engineers which they put to very great use in the study of basic atomic properties. Then, from these quite academic investigations, suddenly arose the idea of inverted energy level populations, with the possibility of emission of radiation; the development of the maser and laser, as fascinating devices of immense practical application followed very rapidly. It can truly be said that the pure scientists have paid back to the microwave engineers the debt which they owed.

No doubt the laser and related devices themselves will be found to have immense applications in the field of pure science. Some fascinating possibilities are already opening up in this connection, since it should now be possible literally to shake molecules to pieces with the high-energy densities that are available. This, in turn, will no doubt prove to have great practical applications. The main theme which can clearly be read into this whole field of scientific development is that science progresses most rapidly and most fruitfully when the pure scientist with his academic pursuits works hand in glove with the applied engineer and his practical problems, and the two together thus share in both the problems and prizes of each other's work.

INDEX

Absorption, 34
   resonance, 35
   spectrum, 44
Aerial, 107
Aether, 35
Amplitude, 16
Aperture synthesis, 61
Atmosphere, 56
Atom, 29

Bradley, 12
Bragg, 81
Bragg's Law, 85
Broglie, de, 26

Cavity resonator, 113
Coherent radiation, 139
Colour, 5
Compression wave, 7
Continuous spectrum, 35, 43

Diffraction, 17
   grating, 20
Doppler effect, 20

Einstein, 36
Electromagnetic spectrum, 1, 17, 126
   wave, 15
Electron resonance, 29
   spin, 34
Energy level, 32
Excited atom, 29

Faraday, 13
Fizeau, 12
Fourier transform, 81
Frequency, 16
Fresnel, 10

Galaxy, 38
Galileo, 12
Gamma radiation, 1
Grimaldi, 5
Gunn effect, 119

Harmonics, 30
Heat radiation, 23
Heisenberg, 28
Herschel, 38
Hertz, 14
Hey, 57
Hooke, 5
Huygens, 66

Incoherent radiation, 139
Infra-red, 2
Interference, 8, 10
Interstellar dust, 45
Intrinsic luminosity, 47
Ionosphere, 56

Jansky, 57

Kirchhoff, 18
Klystron oscillator, 114

Laser, 141
Laue, 80
Light wave, 1
Line of force, 13
Longitudinal wave, 6

Maser, 133
Matter wave, 28
Maxwell, 14
Michelson and Morley, 38
Microwave, 103
    spectroscopy, 127
Millimetre wave, 118
Mode of propagation, 112

Newton, 5

Oscillator, 106

Photo-electric effect, 22
Photon, 25
Planck, 24
Polarisation, 6
Polaroid, 8
Pulsar, 69

Quantum theory, 24
Quasar, 68

Radar, 120
Radio galaxy, 67
Radio telescope, 58
Radioastronomy, 56

Red shift, 54
Relativity, 36
Römer, 12
Röntgen, 74

Secondary waves, 8
Slow-wave, 117
Spectrograph, 42
Spectrum, 29
Speed of light, 12
Starlight, 40
Statistical mechanics, 23
Structure of matter, 75
Synchrotron radiation, 67

Transverse wave, 7

Ultra-violet, 1
    catastrophe, 23
Uncertainty Principle, 28
Universe, 69

Wave frequency, 16
    velocity, 16
Waveguide, 111
Wavelength, 16
White light, 35

X-rays, 2
    diffraction, 74
    generation, 33

Young, 8